Elements of Quantum Mechanics of Infinite Systems

International School for Advanced Studies
Lecture Series No. 3

ELEMENTS OF QUANTUM MECHANICS OF INFINITE SYSTEMS

Lecture notes by F Strocchi

World Scientific
Singapore • Philadelphia

Published by

World Scientific Publishing Co. Pte. Ltd.
5 Toh Tuck Link, Singapore 596224
USA office: 27 Warren Street, Suite 401-402, Hackensack, NJ 07601
UK office: 57 Shelton Street, Covent Garden, London WC2H 9HE

British Library Cataloguing-in-Publication Data
A catalogue record for this book is available from the British Library.

ELEMENTS OF QUANTUM MECHANICS OF INFINITE SYSTEMS

ISBN-13 978-9971-978-91-4
ISBN-10 9971-978-91-1
ISBN-13 978-9971-978-92-1 (pbk)
ISBN-10 9971-978-92-X (pbk)

PREFACE

These notes arose from lectures at the Scuola Normale Superiore (Pisa) and at the International School for Advanced Studies (Trieste) in 1977-78, 1981-82, 1982-83. The lectures were addressed to students of the last year of undergraduate studies or of the first year of graduate studies. The course was planned to discuss some of those basic features of quantum mechanics of systems with infinite degrees of freedom (QM_∞) like collective phenomena, spontaneous symmetry breaking, etc. that in the author's opinion should be part of the common education of every theoretical physics student. No pretention of completeness is made about the subject covered in these lectures. The present notes are only meant to serve as an introduction to the problems and results of QM_∞. The mathematical precision has been reduced to the minimum in order to communicate the main ideas to a larger audience, including people not mathematically minded. The hope is that once the basic structures are known each student may eventually implement the arguments by mathematical rigour without much difficulty, according to his taste, following for example the references as a guide.

The main motivation for writing down the lecture notes was to help the students who apparently found some difficulty in finding an accessible and compact exposition of the material in standard textbooks.

CONTENTS

Elements of Quantum Mechanics of Infinite Systems

INTRODUCTION

In its original formulation, Quantum Mechanics (QM) was essentially dealing with systems with a finite number of degrees of freedom. Invented to explain the atomic levels, QM has been able to explain very different phenomena and its level of completeness and consistency, even from a rigorous point of view, is very satisfactory. There are, however, physical systems whose theoretical description requires infinite degrees of freedom and the treatment of quantum phenomena associated to them requires an extension of the original structure of QM. Such a formulation goes under the various names of QM of systems with infinite degrees of freedom, second quantization, field theory, many-body theory etc. according to the branches of physics to which it has been applied. Contrary to the case of ordinary QM, the QM of systems with infinite degrees of freedom (QM_∞) still presents unsolved questions of principle and its foundations are still the object of investigations and research. Most of what is known about that theory is based on the perturbative expansion and the recently proven triviality of the ϕ^4 theory (in four space-time dimensions) shows how misleading the perturbation expansion may be. Nevertheless, in many of the fields in which the ideas and methods of QM_∞ have been used, the success has beeen so remarkable (often beyond the theoretical expectations) that it is not unreasonable to regard it as a trustable theory, whose physical relevance cannot be denied.

The historical motivations for extending ordinary QM to systems with infinite degrees of freedom mainly came from the problem of combining QM and (special) Relativity. The need of describing interactions (or forces) for relativistic systems no longer as "actions at a distance", but as contact actions, mediated by "fields", necessarily leads to considering systems with infinite degrees of freedom. The well-known classical example is the expansion of the electromagnetic field into infinite normal modes. High energy physics (in particular elementary particle physics) is therefore the physics of systems with infinite degrees of freedom and the theoretical quantum description of them goes under the name of quantum field theory. Its successful applications go beyond the realm of high energy physics, like hyperfine structure of atomic levels (Lamb shift), nuclear physics, etc.

The theoretical description of macroscopic systems in terms of forces between their "elementary" constituents (like atoms, electrons, molecules etc.) also involves systems with infinite degrees of freedom. The essential features of macroscopic bodies is that of consisting of a very large number N of particles and that their physical description in general involves "intensive" or "thermodynamical" properties of such systems (like particle density, mean energy per particle, etc.), i.e. properties for which effects of order 1/N and/or 1/V can be neglected, provided the density $n = N/V$ is kept finite. The above physical properties are therefore essentially the same as for the limiting situation of infinite degrees of freedom ($N \to \infty$) in an infinite volume ($V \to \infty$). This limiting case (also called thermodynamical limit) is at the basis of theories like Statistical Mechanics and Thermodynamics it not only leads to great technical simplifications, like the exploitation of Euclidean invariance, the neglection of surface or finite volume effects etc., but it also allows a simple treatment of concepts like temperature, collective excitations, quasi particles, condensation etc., all of which involve the $N \to \infty$ limit and would hardly be treatable otherwise. The quantum treatment of such branch of physics goes under the name of Many Body Theory. The fields in which many body theory has been successfully applied run from nuclear physics to superconductivity, superfluidity, plasma physics, electron structure of metals, ferromagnatism, etc.

The above remarks should make it clear that QM_∞ should be regarded as a very useful theory especially for its interdisciplinary aspects, as relevant for physics as ordinary QM. As mentioned before, many questions of principle for QM_∞ are still unsolved, especially from a non-perturbative point of view, but the main features and structure of the theory are becoming more and more established. In particular important phenomena like collective phenomena, spontaneous symmetry breaking, which appear to play such a crucial role in modern theoretical physics, have no counterpart in ordinary QM. So that what was historically viewed as the origin of the many difficulties of QM_∞, in comparison with ordinary QM, is now emerging as the basis for (partly) unexpected and welcome structures, which appear as common features of very different physical phenomena.

The aim of these lectures is to emphasize those general features of QM_∞ which have an interdisciplinary interest and which are strictly related to the foundations of the theory itself. As much as possible the discussion will not rely on perturbative expansions and non-perturbative effects will actually be emphasized.

PART A – MANY PARTICLE SYSTEMS AND ELEMENTARY EXCITATIONS

I MANY PARTICLE SYSTEMS

1.1. Canonical variables and their representation

Historically QM_∞ was introduced in connection with the problems of relativistic wave equations(*) but the trick of replacing wave functions with field operators somewhat obscures the simple unavoidable features of any QM of systems with infinite degrees of freedom.

The first step in the description of an infinite system is essentially a kinematical problem, namely the identification of a suitable set of canonical variables $\{q_i, p_i\}$, $i = 1, \ldots$, just as in the case of a classical system. The canonicity condition amounts to the fulfillment of the canonical commutation relations (CCR)

$$[q_i, p_j] = i\,\delta_{ij} \tag{1.1}$$

(In the classical case we would have had Poisson brackets instead of commutators). As in the classical case, the canonicity conditions are independent from the dynamics, i.e. from the specification of the Hamiltonian, and have essentially a kinematical or algebraic content. One is naturally led to consider the algebra $\mathcal{A}$ of the canonical variables as the algebra generated by the q's and p's through multiplications and sums, with eq.(1.1)

(*) See e.g. N.N. Bogoliubov and D.V. Shirkov, Introduction to the Theory of Quantized Fields, Interscience, New York, 1957, Chapt. I, II.

as algebraic relations within $\mathcal{A}$. In principle, the canonical variables must provide a complete description of the system (at a given time) and therefore any physical quantity A should be expressible as a function A = A (q,p) of the canonical variables.

Remark At this point two technical remarks are necessary. First eqs. (1.1) imply that when represented as linear operators in a Hilbert space, the p's and q's cannot be both bounded operators, therefore to avoid domain questions it is better to use "bounded" functions of the p's and q's to generate the algebra $\mathcal{A}$. This is done by using the so-called Weyl operators

$$U_i(\alpha) = e^{iq_i\alpha} \quad , \qquad V_j(\beta) = e^{ip_j\beta} \quad , \quad \alpha, \ \beta \in \mathbb{R} \qquad (1.2)$$

in terms of which the CCR read

$$U_i V_j = V_j U_i \quad e^{i \ \delta_{ij} \ \alpha\beta} \qquad (1.1')$$

The reality (or hermiticity) of the p's and q's naturally defines a * operation in $\mathcal{A}$: $U_i^*(\alpha) = U_i(-\alpha)$, $V_j^*(\beta) = V_j(-\beta)$. Technically $\mathcal{A}$ is thus a * algebra. The second point is that with the above prescription $\mathcal{A}$ is a polynomial algebra generated by the U's and V's and to construct non-polynomial functions one needs a concept of convergence. This is done by assigning a norm to each element A of $\mathcal{A}$ such that $||A^*A|| = ||A||^2$ and to complete $\mathcal{A}$ with respect to this norm. The so completed algebra, which we still denote by $\mathcal{A}$, is technically a C*-algebra.

Example 1. As an example of a choice of canonical variables for a continuum infinite system described by the real field $\phi(x)$ one can fix a complete set of real orthonormal functions $f_i(x) \in L^2(R^3)$ and define

$$q_i(t) = \int d^3x \; f_i(x) \quad \phi(\vec{x},t)$$
$$p_i(t) = \int d^3x \; f_i(x) \quad \dot{\phi}(\vec{x},t)$$

Clearly the states of the system at a time t is completely specified by the q's and p's since $\phi(x,t)$, $\dot{\phi}(x,t)$ can be recovered from them.

Once the canonical variables (i.e. the algebra) have been specified, in order to determine the time evolution of the states of the system, one has to fix the represnetation of the canonical variables as operators in a Hilbert space H. In fact only after the representation of the q's and p's has been chosen, the Hamiltonian (or better the time translation operator $U(t) = \exp i\, Ht$) can be written as a well defined operator in H, and the dynamical problem is reduced to an eigenvalues problem in H.

The important point is that for systems with a finite number of degrees of freedom ($N < \infty$), the choice of the representation has a purely kinematical character since, as proved by Von Neumann(*), for $N < \infty$ all the irreducible representations of the algebra $\mathcal{A}$ are unitary equivalent. This means that one can go from one representation to another by a unitary transformation (i.e. by the quantum analog of canonical transformations); clearly the physical description of the system will not depend on the choice of the representation, which is therefore reduced to a pure matter of convenience.

(*) J. Von Neumann, Math.Ann. 104, 570 (1931)
For a detailed exposition of these results see e.g. C. Putnam, Commutation Properties of Hilbert Space Operators, Springer 1967, Ch. IV.

For the $N = \infty$ case, the situation changes drastically, since any two irreducible representations of $\mathcal{A}$ are in general unitarily inequivalent. Different inequivalent representations will in general give rise to different physical pictures with different physical implications. As we will see in the following sections, the Hamiltonian H(q,p) will in general turn out to be an acceptable Hamiltonian (e.g. bounded below, positive, etc.) only for a particular representation (and for those unitarily equivalent to it). Otherwise the Hamiltonian will have a pathological spectrum, for example $H = \infty$. As a result for $N = \infty$, the choice of the representation is no longer a kinematical fact, but it is strongly linked to the dynamics. An incorrect choice of the representation is in fact at the origin of the divergence problem which afflicts the perturbative expansions of quantum field theory and of many body theory. This explains why QM_∞ is much more difficult than ordinary QM , $(N < \infty)$, and also why QM_∞ exhibits such interesting phenomena like collective effects, spontaneous symmetry breaking etc. which have no counterpart in the $N < \infty$ case.

As we will see, to solve the arbitrariness involved in the choice of the representation one must in general specify additional requirements, related to the physical properties of the system; typically information about the structure of the ground state are needed (a higly dynamical problem!). From a mathematical point of view a basic problem of QM_∞ is the <u>representation problem.</u>

1.2 Ground state and elementary excitations. Fock representation

We will discuss a natural choice of canonical varaibles, which are very convenient for the description of infinite systems. Such canonical variables have a very simple and physical justification if the states of the system are described in terms of occupation numbers (see below), but they have a much more general right of existence.

The basic idea underlying the occupation number (or Fock) representation is that the states of the system (at a given time) are analyzed by making reference to the ground state and by specifying the number and the type of elementary excitations which characterize a given state, with respect to the ground state. Here and in the following by elementary excitation or particle we mean the generic excitation of the system like e.g. a sound wave in a crystal, a collective excitation in a plasma or in a nucleus, an elementary particle in field theory etc.The elementary excitation is completely identified once a complete set of observables $\{\alpha\}$ is specified characterizing the physical properties of it. For example, for an elementary particle a complete set of one-particle observables may be the momentum $\vec{k}$, the spin s_z, the charge and possibly other internal quantum numbers. The one-particle states are then labelled by the set of eigenvalues α':

$$\alpha|\alpha'\rangle = \alpha'|\alpha'\rangle$$

and the many particle states will be analized in term of single particle quantum numbers, for example

$$|\alpha_1', \alpha_2' \ldots \alpha_n'\rangle \equiv |\alpha_1'\rangle |\alpha_2'\rangle \ldots |\alpha_n'\rangle$$

is an n-particle state specified by the particle 1 being in the state $|\alpha_1'\rangle$, the particle 2 in $|\alpha_2'\rangle$ etc. The rational at the basis of this representation is the following: i) the representation is the quantum analog of that used in classical statistical mechanics when the microscopic states of the system are described in terms of occupation numbers of cells in phase space, ii) the representation is very suitable for scattering states since the description in terms of wave functions of the single particles is very close to the experimental situation, iii) above all, this representation is useful for the description of physical processes, in which the number of particles or of elementary excitations of the system is not a constant of motion (creation and annihilation can take place), since a transition from an n-particle state to an n + 1 particle state does not require a redefinition of the observables relative to the n particles, as it would be the case if global observables like the total energy or the total angular momentum are used to describe the n-particle state. Here, to describe a process in which a new particle is created, it is enough to specify the quantum number of the additional particle, without changing the description of the other n.

The next step for obtaining the Fock representation is to take particle identities into account, by specifying only the number n_i of identical particles which are in a given state $|\alpha_i'\rangle$. For identical particles the base states are then specified in the following way. One chooses once and for all the one-particle states and orders (*) them in some way:

(*) For simplicity we consider observables with discrete spectrum. The difficulties involved in the case of observables with continuum spectrum (like position, momentum, energy) are essentially technical and mathematical. With an abuse of language, we will sometimes introduce a position and momentum eigenstates ($|\vec{x}\rangle, |\vec{k}\rangle$), leaving to the reader the task of making the arguments rigorous, by the introduction of normalized wave packets

$|\alpha_1'>$, $|\alpha_2'>$, ... The n-particle state is then identified by the sequence $\{n_1, n_2, \ldots\}$, where n_i specifies the number of particles in the state $|\alpha_i'>$. Clearly $\sum n_i = n$ and only a finite number of n_i 's are different from zero. Furthermore, if the particles or elementary excitations are <u>bosons</u> there is no restriction on the numbers n_i, whereas for <u>fermions</u>, the Pauli principle requires $n_i = 0$ or $n_i = 1$. The n_i are called occupation numbers and the basic vectors are denoted by $\Psi^{(n)}_{n_1, n_2 \ldots}$ or by $|n; n_1, n_2 \ldots >$, with normalization

$$< n'; n_1', n_2' \ldots | n; n_1, n_2 \ldots > = \delta_{n_1 n_1'} \delta_{n_2 n_2'} \cdots$$

The Hilbert space of the states of the system is the space generated by the above vectors, when n = 0, 1, 2, ..., and for any two vectors $\Phi, \Psi \in H$ the scalar product is

$$< \Phi, \Psi > = \sum_{n=0}^{\infty} < \Phi^{(n)}, \Psi^{(n)} > ,$$

with $\Phi^{(n)}$, $\Psi^{(n)}$ the projections on the n-particle subspace $H^{(n)}$.

In the Fock representation, it is very natural to introduce the so-called annihilation and creation operators defined as follows. For each single particle state α_i the annihilation operator a_i is defined by:

$$a_i | n_1, \ldots, n_i, \ldots > = \sqrt{n_i} \, | n_1, \ldots, n_i - 1, \ldots > , \qquad (1.3)$$

for bosons and by

$$a_i | n_1, \ldots n_i, \ldots > = (-1)^{\theta_i} n_i | n_1, \ldots n_i - 1, \ldots > \qquad (1.4)$$

with $\theta_i = \sum_{k=1}^{i-1} n_k$, for fermions. The choice of the phases and of the normalizations will be justified below.

In a similar way one introduces the creation operators

$$b_i \mid n_1, \ldots, n_i, \ldots > = \sqrt{n_i + 1} \, |n_1, \ldots, n_i + 1, \ldots > \tag{1.5}$$

$$b_i \mid n_1, \ldots, n_i, \ldots = (-1)^{\theta i} (1 - n_i) \mid n_1, \ldots, n_i + 1, \ldots > \tag{1.6}$$

for bosons and fermions respectively. The factors $(1 - n_i)$ and n_i for the fermion case are there to authomatically take care of Pauli principle: one cannot create (or destroy) a particle in the i-th state if such state is already occupied $(n_i = 1)$ (or unoccupied, $n_i = 0$). It is very simple to see that for bosons

$$[a_i, a_j] = 0 \qquad [b_i, b_j] = 0 \tag{1.7}$$

whereas for fermions

$$\{a_i, a_j\} = a_i a_j + a_j a_i = 0, \qquad \{b_i, b_j\} = 0 \tag{1.8}$$

Furthermore with the above choice of phases, b_i turns out to be the adjoint of a

$$b_i = a_i^*$$

and the following equations hold

$$[a_i, a_j^*] = \delta_{ij} \tag{1.7'}$$

for bosons and

$$\{a_i, a_j^*\} = \delta_{ij} \tag{1.8'}$$

for fermions. The relations (1.7) (1.7') are called the canonical commutation relations (CCR) for the bosonic creation and annihilation operators, and the (1.8)(1.8') are called the canonical anticommutation relations (CAR).

The number of particles in the i-th state can be easily expressed in terms of the a_i , a_i^* :

$$N_i = a_i^* a_i$$

since

$$a_i^* a_i \mid \ldots n_i \ldots > = n_i \mid \ldots n_i \ldots >$$

The total number of particles is $N = \sum_{i=1}^{\infty} N_i$.
It follows from the CAR that (for fermions)

$$a_i^2 = 0 = a_i^{*2}$$

and

$$N_i^2 = a_i^* a_i a_i^* a_i = a_i^* (1 - a_i^* a_i) a_i = a_i^* a_i = N_i$$

i.e. the eigenvalues of N_i are 0 or 1, in agreement with Pauli principle.

Similarly one can easily express the total free Hamiltonian in terms of a, a^*. If the one-particle states have definite energy ω_i, then

$$H = \Sigma \omega_i a_i^* a_i = \Sigma \omega_i N_i , \tag{1.9}$$

with obvious physical meaning. From the CCR (or CAR) one gets

$$[H , a_i^*] = \omega_i a_i^* , \quad [H, a_i] = - \omega_i a_i$$

and therefore the time evolution of a_i is

$$a_i(t) = e^{-iHt} a_i e^{iHt} = e^{-i\omega_i t} a_i$$

The operators a_i , a_i^* have been introduced by making reference to the occupation number representation, but clearly they can be regarded as canonical variables introduced at a purely kinematical level, without reference to a given representation. For example, they may be introduced in terms of the q's and p's by the formula

$$a_j = \frac{1}{\sqrt{2}} (q_j + i\, p_j)$$

or by the properties of diagonalizing the Hamiltonian (in the sense of eq. (1.9), see sect. 1.3). As before, one can introduce the algebra $\mathcal{A}$ generated by a, a^* (or better by the bounded functions of a, a^*) and one can investigate the possible irreducible representations of $\mathcal{A}$ (*)

The occupation number (or Fock) representation (†) can be characterized by the existence of a state Ψ_o, called the ground state, such that

$$a_i \, \Psi_o = 0 \qquad\qquad \forall i \tag{1.10}$$

In fact, if this condition is satisfied one can easily construct the n-particle state $|n_1, n_2 \ldots>$ by successive applications of a^* to Ψ_o

$$| n_1 , n_2 \ldots > = \frac{1}{\sqrt{n_1!\, n_2! \ldots}} \underbrace{a_1^* \ldots a_1^*}_{n_1} \; \underbrace{a_2^* \ldots a_2^*}_{n_2} \ldots | \Psi_o > ,$$

since, by eq. (2.9), $N_i \, \Psi_o = 0$ and

$$N_i \; a_1^* \ldots a_2^* \ldots \Psi_o = [N_i , a_1^* \ldots a_2^* \ldots] \Psi_o = n_i \; a_1^* \ldots a_2^* \ldots \Psi_o .$$

(The normalization factor $(n_1!\, n_2! \ldots)^{-1/2}$ is related to the choice of normalizations in eqs. (1.3) - (1.6)).

The Fock representation is also characterized by the property that

---

(*) The unitary equivalence of all irreducible representations for $N < \infty$, is given by Von Neumann theorem for the bosonic case and by Jordan and Wigner (Zeit. F. Phys. 97, 650 (1928)) for the fermionic case.

(†) In the following, the Fock representation will be understood to be irreducible.

the number of particles N is a well defined operator with a discrete (positive) spectrum. In fact, if N is well defined and Ψ_λ is an eigenstate with eigenvalue λ , then

$$N a_i \Psi_\lambda = \lambda a_i \Psi_\lambda + [N, a_i] \Psi_\lambda = (\lambda - 1) a_i \Psi_\lambda . \qquad (1.11)$$

If $\lambda > 0$ then

$$0 < \lambda \| \Psi_\lambda \|^2 = < \Psi_\lambda , N \Psi_\lambda > = \sum_i \| a_i \Psi_\lambda \|^2$$

and therefore there must be at least an index i for which $a_i \Psi_\lambda \neq 0$ and eq. (1.11) holds. Thus if $\lambda > 0$ is an eigenvalue of N also $\lambda - 1$ is an aigenvalue of N. Since the spectrum of N is positive, λ must be an integer and $\lambda = 0$ is a point in the spectrum of N. Clearly, the corresponding eigenvector Ψ_0 satisfies the Fock condition (1.10) :

$$0 = < \Psi_0 , N \Psi_0 > = \sum_i \| a_i \Psi_0 \|^2 \Rightarrow a_i \Psi_0 = 0 \quad \forall i .$$

The above argument may be used to show that for a finite number of degrees of freedom, in <u>any</u> representation of the CCR (or CAR) one can always construct a ground state satisfying the Fock condition. In fact, by definition of representation, there is a common dense domain D such that $a_i D \subset D$, $a_i^* D \subset D$, and since N is a <u>finite</u> sum of positive operators $a_i^* a_i$, N is well defined on D. By applying essentially the above argument, eq. (1.11), to the spectrum of N one shows that if λ is a positive point of the spectrum, then also $\lambda - 1$ is a point of the spectrum. The positivity of N then requires λ to be a positive integer and 0 to be an isolated point of the spectrum of N.

Now all the irreducible representations with a Fock state Ψ_o are unitarily equivalent, since for any polynomial P(a,a*) of a,a* the correspondence

$$P(a,a^*)\Psi_o \rightarrow P(a', a'^*)\Psi'_o$$

preserves the scalar product and it is therefore described by a unitary operator. In this way, one gets a proof of Von Neumann theorem for $N < \infty$.

Problem.(*) Show that if the Hamiltonian has the form

$$H_p \equiv \Sigma\omega_k a^*_k a_k$$

with $\omega_k > \omega_o > 0$ (mass gap) than any representation, in which H is a well (densely) defined operator, is unitarily equivalent to the Fock representation. (Hint: since $\omega_k / \omega_o > 1$, one has $N < \frac{1}{\omega_o} H$ and therefore whenever H is well defined also N is well defined). This shows that in the presence of interactions the splitting $H = H_o + H_{int}$ has a meaning, i.e. H_o is well defined (a necessary condition for perturbation theory) only if the representation for H is unitarily equivalent to the Fock representation.

The physical meaning of the Fock property (1.10) is that in QM_∞ , as long as one can count the number of elementary excitations associated to certain degrees of freedom, the representation of the canonical variables relative to such degrees of freedom is (unitarily equivalent to) a Fock representation. On the other side, if the dynamics of some degrees of freedom is such that one can no longer count the number of the corresponding elementary

(*) H.J. Borchers, R. Haag and B. Schoer, Nuovo Cimento 29 (1963) 148.

excitations (for example because such degrees of freedom get "frozen" or "confined" as a result of the interaction, or give rise to a condensation with any state containing an infinite number of elementary excitations etc.) the representation cannot be a Fock one. We will see many examples of these very interesting phenomena. Quite generally, if a_i , a_i^* are the annihilation and creation operators of the elementary excitations of the non-interacting system, the ground state of the interacting system will not satisfy the Fock condition for the a_i's. Form a physical point of view, this means that in general the interaction induces a redefinition of the degrees of freedom of the non-interacting system (collective effect). In this case the choice of the the representation of the CCR (or CAR) is a non-trivial problem(*). Clearly all these phenomena have no counterpart in ordinary QM.

(*) For an excellent review of this problem see A.S. Wightman, in Proc.Int. Conf. on Particle and Fields, Rochester 1967 , C. Hagen et al. eds.

1.3 Quantum vibrations in a crystal. Phonons

To simplify the discussion we start with a one dimensional chain of atoms with

L = length of the chain

N = number of atoms

a = lattice distance (Na = L)

Boundary effects are taken care of by imposing periodic boundary conditions. This also guarantees invariance under discrete translations of step a.

As a first approximation we will consider next neighbouring interaction by an harmonic potential with elastic constant λ . If q_i, p_i denote the deviation from the equilibrium position and the momentum of the i-th atom, the Hamiltonian takes the form

$$H = \frac{1}{2} \sum_i \left[\frac{p_i^2}{2m} + \lambda (q_i - q_{i+1})^2 \right] \tag{1.12}$$

(m = atomic mass).

According to the rules of QM the canonical variables q,p are required to satisfy the CCR

$$[q_i , p_j] = i \, \delta_{ij} \tag{1.13}$$

As in the case of single harmonic oscillator one can diagonalize the Hamiltonian by a normal mode exp sion of the q's and p's:

$$q_\ell = \frac{1}{\sqrt{2N}} \sum_k (m\omega_k)^{-\frac{1}{2}} [a_k e^{ik\ell a} + a_k^* e^{-ik\ell a}] \tag{1.14}$$

$$p_\ell = \frac{-i}{\sqrt{2N}} \sum_k (m\omega_k)^{\frac{1}{2}} [a_k e^{ik\ell a} - a_k^* e^{-ik\ell a}] \tag{1.14'}$$

where

$$\omega_k = \left[\frac{2\lambda}{m} (1 - \cos k a) \right]^{\frac{1}{2}}$$

is the classical frequency of the normal mode. For small wave vectors k ($ak \ll 1$ or $k \ll \frac{1}{a}$) we get the linear dispersion relation

$$\omega_k = \sqrt{\lambda/m}\, a\, k \equiv \omega_o\, a\, k$$

The periodicity conditions

$$q_{i+N} = q_i$$

then yield

$$k = \frac{2\pi n}{Na} = \frac{2\pi n}{L}, \quad n = 0, \pm 1, \pm 2 \,..$$

Finally since q_ℓ , p_ℓ do not change if $k \to k + (2\pi/a)$, the variable k is only defined modulo $2\pi/a$ and by convention will be chosen to lie in the interval $[-\pi/a, \pi/a]$. Hence

$$k = \frac{2\pi}{a}\frac{n}{N}, \quad n = 0, \pm 1, \ldots \tfrac{1}{2} N \qquad (N \text{ even})$$
$$n = 0, \pm 1, \ldots \pm\tfrac{1}{2}(N-1) \qquad (N \text{ odd}) \tag{1.15}$$

From eqs.(1.14) and (1.14') it follows that a_k , a_k^* obey the CCR

$$[a_k, a_{k'}] = 0 = [a_k^*, a_{k'}^*], \qquad [a_k, a_{k'}^*] = \delta_{k,k'}$$

and the Hamiltonian is diagonal in the variables a, a^*

$$H = \sum_k \omega_k (a_k^* a_k + \tfrac{1}{2})$$

The elementary excitations (or quantum sound waves) are then characterized by the wave vector k and by energy ω_k . It should be stressed that k is not a true momentum (see eq.(1.15); it is called "pseudo-momentum" and the corresponding conservation law involves also the momentum that the lattice can exchange as a whole. For example, for an ion-phonon interaction one has

$$\hbar q = \hbar k + \hbar K_n$$

where $\hbar q$ is the momentum given by the ion, $\hbar k$ is the phonon momentum and $K_n = \pm n 2\pi/a$ is the momentum received by the lattice as a whole, also called the wave vector of the reciprocal lattice. The above conservation law follows from the invariance under (discrete) translations of multiples of a.

For the three dimensional case one proceeds in a similar way, with the difference that for each k the elastic constant is actually a tensor $K_{ij}(k)$. The normal modes are then found by diagonalizing the potential energy

$$\tfrac{1}{2} K_{ij}(\vec{k})\, q_i(\vec{k})\, q_j(\vec{k})$$

i.e. by determining the principal axes $n_1(\vec{k})$, $\vec{n}_2(\vec{k})$, $\vec{n}_3(\vec{k})$ of $K_{ij}(\vec{k})$ for each k. The creation and annihilation operators $a^*_\alpha(\vec{k})$, $a_\alpha(\vec{k})$ for the mode along the α-th axis are then introduced through the equation

$$\vec{q}_{\vec{\ell}} = \frac{1}{\sqrt{2N}} \sum_{k,\alpha} [\, m\omega(k,\alpha)\,]^{-\frac{1}{2}}\, \vec{\epsilon}_\alpha(\vec{k})\, [\, a_\alpha(\vec{k}) e^{i\vec{k}.\vec{\ell} a} + a^*_\alpha(\vec{k})\, \ell^{-i\vec{k}.\vec{\ell} a}\,] ,$$

where $\vec{\epsilon}_\alpha(\vec{k}) = \vec{n}_\alpha(\vec{k})$ denote the polarization vectors. They can be decomposed along transverse, $\vec{e}_{T\alpha}(\vec{k})$, and longitudinal, $\vec{e}_{L\alpha}(\vec{k})$, polarizations

$$\vec{\epsilon}_\alpha(\vec{k}) = \vec{e}_{T\alpha} + \vec{e}_L , \qquad \vec{k} \cdot \vec{e}_{T\alpha} = 0, \quad \vec{k} \cdot \vec{e}_{L\alpha} = |\vec{k}| .$$

The Hamiltonian then takes the form

$$H = \tfrac{1}{2} \sum_{k,\alpha} \omega(\vec{k},\alpha)\, [\, a^*_\alpha(k)\, a_\alpha(k) + a_\alpha(k)\, a^*_\alpha(k)\,] \tag{1.16}$$

To discuss the general features of the spectrum $\omega(k,\alpha)$ it is convenient to consider simplified models. We start by considering elastic waves with definite wave vector k moving along such directions that only the transverse mode (or the longitudinal mode) propagates. The atom displacements on planes which are perpendicular (or parallel) to the direction of propagation are then in phase and the elastic wave can be described by just one co-ordinate. One is then reduced to the one-dimensional case. If all the atoms are identical one recovers exactly the linear chain discussed before with $\omega(k) = c\,k$ for small k. Such waves are the typical acoustic waves in solids with propagation speed, $d\omega/dk$, given by the sound velocity.

Another interesting case is given by a linear chain consisting of two kind of atoms of mass m_1 and m_2, respectively, which alternate at distance a/2. In this case one gets two types of normal modes, with frequency ω_1 and ω_2 :

$$\omega_1^2 = \frac{K}{\mu}\left\{1 - \left[1 + \frac{4\,m_1\,m_2}{(m_1+m_2)^2}\sin^2\left(\frac{ka}{2}\right)\right]^{1/2}\right\},$$

$$\omega_2^2 = \frac{K}{\mu}\left\{1 + \left[1 + \frac{4\,m_1\,m_2}{(m_1+m_2)^2}\sin^2\left(\frac{ka}{2}\right)\right]^{1/2}\right\},$$

where K is the elastic constant and μ is the reduced mass

$$\mu = m_1 m_2/(m_1 + m_2).$$

The ω_1 mode corresponds to a vibration in which the two kinds of atoms move in phase; it is the analog of the elastic wave of a linear chain consisting of identical atoms, and in fact, it is characterized by

$$\omega_1(k) \sim c\,k \qquad \text{for small k.}$$

The ω_2 mode is characterized by

$$\omega_2(k) \sim (2K/\mu)^{1/2}$$

i.e. independent of k for small k. It corresponds to a motion in which the two atoms move out of phase. In each cell, containing a pair of atoms m_1 and m_2, the pair vibrates as if the other atoms were not present and if the two atoms have opposite charge, as it is the case for ionic or polar crystals, the vibration gives rise to an optically active dipole (optical mode).

The two modes mentioned above are described by normal coordinates which for low k correspond to the center of mass coordinate $(Q_{2\ell} \simeq (m_1 q_{2\ell} + m_2 q_{2\ell+1})/(m_1+m_2))$ and to the relative coordinate $(Q_{2\ell} = q_{2\ell+1} - q_{2\ell})$.(*)

1.4 Field operators

QM_∞ was originally deviced to solve the problems of relativistic wave mechanics and as such it was based on the concept of quantum fields. In the framework discussed so far a field operator $\psi(x)$ can be introduced by the transformation

$$\psi(\vec{x}) = \sum_i f^s_{\alpha_i}(\vec{x})\ a_i \tag{1.17}$$

where $f_{\alpha_i}(\vec{x}) \equiv \langle \vec{x},s \mid \alpha_i\rangle$ are the wave functions of the one particle states $\mid \alpha_i \rangle$ and s denotes the spin variable and possibly other quantum numbers, like charge. (For brevity in the following the index s will often be omitted). Similarly, in momentum space one can introduce the operators

$$a(\vec{k},s) = \sum_i f^s_{\alpha_i}(\vec{k}) a_i = \sum_i \langle \vec{k},s \mid \alpha_i\rangle\ a_i \tag{1.18}$$

The new operators (1.17)(1.18) essentially describe annihilation of a particle at position $\vec{x}$ (or with sharp momentum $\vec{k}$). Strictly speaking, this is an abuse of language since eigenstates of position (or of momentum) do not exist and

(*) For a more precise and detailed treatment see e.g. R. Kubo and T. Nagamiya Solid State Physics, Mc Graw-Hill Book Co. 1968, Part I, Chap.3.

one should rather think in terms of one particle states corresponding to very narrow wave packets $h_j(\vec{x})$ with mean position $\bar{\vec{x}}$ (or with momentum peaked at $\bar{\vec{k}}$). This would amount to average e.g. the field operator $\psi(\vec{x})$ with the wave functions $h_j(\vec{x})$.

One can easily see that the CCR's (or CAR), eqs.(1.7), (1.7'), (1.8) (1.8'), imply the following (anti) commutation relations

$$\begin{aligned} [\psi(\vec{x},s),\ \psi(\vec{y},s')]_{\pm} &= 0 = [\psi^*(\vec{x},s),\ \psi^*(\vec{y},s')]_{\pm} \\ [\psi(\vec{x},s),\ \psi(\vec{y},s')]_{\pm} &= \delta(\vec{x}-\vec{y})\delta_{ss'} \end{aligned} \tag{1.19}$$

or

$$[\ a(\vec{k},s),\ \ a^*(\vec{k}',s')]_{\pm} = \delta(\vec{k}-\vec{k})\delta_{ss'},\ \text{etc.}\ , \tag{1.20}$$

where the + denotes the anticommutator for fermions, and the - denotes the commutator for bosons. The occurrence of δ-functions is clearly a consequence of the improper nature of the states $|\vec{x},s>$ or $|\vec{k},s>$.

As before one can introduce a Fock representation based on the one particle states $|\vec{x},s>$ (or $|\vec{k},s>$). An n-particle states will then be denoted by $|\vec{k}_1s_1,\ \vec{k}_2s_2,\ \ldots\ \vec{k}_ns_n>$ and since the index i labelling the states has been replaced by a continuum variable $\vec{k}$ the normalization for the creation and annihilation operators now reads

$$a^*(\vec{k})\ |\vec{k}_1,\ \vec{k}_2,\ldots\ \vec{k}_n> = \sqrt{n+1}\ |\vec{k},\ \vec{k}_1\ \ldots\ \vec{k}_n>$$

$$a\ (\vec{k})\ |\vec{k}_1,\ \vec{k}_2,\ldots\ \vec{k}_n> = \frac{1}{\sqrt{n}}\sum_j \delta(\vec{k}-\vec{k}_j)|\vec{k}_1\ \ldots\hat{\vec{k}}_j\ldots\ \vec{k}_n>$$

(the symbol $\hat{k}_j$ means that this variable has to be omitted) so that

$$|\vec{k}_1\ \ldots\ \vec{k}_n> = \frac{1}{\sqrt{n!}}\ a^*(\vec{k}_1)\ldots a^*(\vec{k}_n)\ |0\ >$$

Exercise. Show that the normalized n-particle state $\Psi_{\{n_1 \ldots\}}$ is described by the following wave function

$$\Psi^{(n)}_{\{n_1,\ldots\}}(x_1,\ldots x_n) = C_{\{n_1\ldots\}} \frac{1}{n!} \sum_P \delta_P P(f_{\alpha_{i_1}}(x_1)\ldots f_{\alpha_{i_n}}(x_n))$$

where the sum is over all the permutations P, $\delta_P = 1$ for bosons, δ_P = parity of the permutation for fermions, P is the permutation operator(*) $PF(x) = F(Px)$ and the normalization constant

$$C_{\{n_1\ldots\}} = \sqrt{\frac{n!}{n_1!\, n_2! \ldots}}$$

for bosons, and

$$C_{\{n_1\ldots\}} = \sqrt{n!}$$

for fermions.

(*) For example if

$$P = \begin{pmatrix} 1 & 2 & \ldots \\ j_1 & j_2 & \ldots \end{pmatrix}$$

one has

$$(PF)_{\alpha_1\ldots\alpha_n}(x_1\ldots x_n) = F_{\alpha_1\ldots\alpha_n}(x_{j_1}\ldots x_{j_n}) = F_{(P^{-1}\alpha)}(x_1\ldots x_n)$$

1.5 Second quantization

To complete the description of an infinite system in terms of elementary excitations we will discuss the representation of physical quantities in terms of creation and annihilation operators.

i) *One-body operators*

Let $A^{(1)}$ be the operator that in ordinary QM describes a single particle observable (like e.g. the momentum, the spin, etc.) and, with a suitable choice of the single particle states, can be represented by a diagonal operator

$$A^{(1)} | \alpha_i > = A_i | \alpha_i > . \tag{1.21}$$

The corresponding operator for the infinite system in which an arbitrarily large number of particles can be excited is given by

$$A = \sum_i A_i a_i^* a_i \tag{1.22}$$

The physical meaning is obvious: A counts the number of particles in the i-th state and multiplies this number by the corresponding eigenvalue A_i. For example, if $A^{(1)}$ describes the kinetic energy of one particle, A describes the total kinetic energy of the infinite system. With the above simple construction (1.21) (1.22), one easily recovers the characteristic features of the so-called second quantization whose historical motivations are rooted in the attempt of building a relativistic wave mechanics. In fact, in terms of the field operators introduced in eq. (1.17) the above operator A becomes

$$A = \int dx\, dx' \sum_i A_i < x' | \alpha_i >< \alpha_i | x > \Psi^*(x')\Psi(x)$$

$$= \int dx\, dx' < x' | A^{(1)} | x > \psi^*(x')\psi(x)$$

For example, if $A^{(1)}$ is the one particle kinetic energy one has

$$<x'|\, E^{(1)}_{kin} |\, x> = - \frac{\Delta}{2m}\, \delta(x - x')$$

and

$$E_{kin} = \int dx \;\; \psi^*(x) \;\; \left(-\frac{\Delta}{2\,m}\right) \psi(x) \qquad (1.23)$$

Similarly for the momentum $\vec{P}$ one has $<x'|\, \vec{P}^{(1)} \,|\, x> = -i\, \vec{\nabla}\delta(x - x')$ and

$$\vec{P} = \int dx \;\; \psi^*(x)\; (-i\vec{\nabla})\, \psi(x)$$

Quite generally the expression for A can be formally obtained from the old wave mechanical expression with the wave function replaced by a field operator (second quantization). As a matter of fact, no further quantization has been made; only the extension to multiparticle systems has been involved. Clearly the ordinary QM is recovered by projecting on a definite n-particle space. For example, in the one-particle subspace the operator $\psi^*(x)\psi(x)$ reduces to

$$<\alpha_k|\;\; \psi^*(x)\psi(x) \;\;|\, \alpha_k> = f^*_{\alpha_k}(x)\; f_{\alpha_k}(x)$$

and the above expressions reduce to the ordinary QM ones.

ii) *Two-body operators*

A two-body observable $A^{(2)}$ describes a property of a pair of particles. With a suitable choice of basis one can write

$$A^{(2)} |\alpha_i\, \alpha_j> = A_{ij}\; |\alpha_i \alpha_j> \; , \;\; (A_{ij} = A_{ji})$$

and the corresponding operator for the infinite system is

$$A = \tfrac{1}{2} \sum_{i \neq j} A_{ij}\; a^*_i\; a^*_j\; a_j\; a_i$$

The factor ½ is there to count each pair only once and the case j = i is excluded in the sum to obtain a genuine two-body operator. In terms of field operators A takes the following form

$$A = \int dx\ dy\ dx'\ dy' < x\ y|\ A^{(2)}|\ x'\ y'> \psi^*(x)\psi^*(y)\psi(y')\psi(x')$$

For example if $A^{(2)}$ is a two-body interaction potential,

$$< xy\ |\ A^{(2)}\ |\ x'y'\ > = \tfrac{1}{2}\, V(x,y)\ (\delta(x - x')\delta(y - y') + \delta(x - y')\delta(y - x'))$$

and for the infinite system one gets

$$H_{int} = \int dx\ dy\ \ V(x,y)\ \psi^*(x)\psi^*(y)\ \psi(y)\psi(x)$$

For translational invariant interactions V(x,y) = V(x - y) and in terms of momentum field operators the above interaction Hamiltonian becomes

$$H_{int} = \int dx\ dp\ dq\ a^*(\vec{k} + \vec{p})a^*(\vec{p} - \vec{q})\ \tilde{V}(\vec{q})\ a(\vec{k})\ a(\vec{p}) \qquad (1.24)$$

The above formula allows a very simple diagrammatic interpretation of the elementary processes induced by H_{int}. To first order, the matrix elements $< f|\mathsf{h}|i>$ of the Hamiltonian density $\mathsf{h} = a^*(\vec{k} + \vec{q})\ a^*(\vec{p} - \vec{q})\ V(\vec{q})\ a(\vec{k})\ a(\vec{p})$ do not vanish only if in the initial state $|i>$ there are two particles with momentum k and p, which are destroyed by $a(\vec{k})\ a(\vec{p})$, and in the final state $|f>$ there are two particles of momentum $\vec{k} + \vec{q}$, $\vec{p} - \vec{q}$, which are created by $a^*(\vec{k} + \vec{p})\ a^*(\vec{p} - \vec{q})$. The strength of the elementary process is governed by $\tilde{V}(\vec{q})$.

1.6 Electromagnetic field. Coherent states

In the original formulation of QM_∞ as quantum field theory the emphasis was on the concept of field and the concept of particle was one of the most interesting consequences; in particular, the quantization of the electromagnetic field leads to the explanation of photons as elementary excitation of the e.m. field. Here, we will follow the opposite route by showing that the field operator for a system of massless particles with spin one and zero charge, obeys the Maxwell equations and therefore it provides a quantization of such equations.

We start by chosing as single particle observables the momentum $\vec{k}$ and the helicity λ. The zero mass implies that the energy $\omega(\vec{k}) = |\vec{k}|$ and the helicity $\lambda = \pm 1$. The single particle (photon) states will then be denoted by $|\vec{k}, \lambda >$ and the annihilation and creation operators by $a(\vec{k},\lambda)$ and $a^*(\vec{k}, \lambda)$. For each $\vec{k}$, we chose two complex polarization vectors $\vec{\varepsilon}(\vec{k}, \lambda)$, $\lambda = \pm 1$, satisfying the transversality condition

$$\vec{k} \cdot \vec{\varepsilon}(k, \lambda) = 0$$

and

$$\vec{\varepsilon}(k, \lambda)^* = \vec{\varepsilon}(-k, \lambda) \quad , \qquad \vec{\varepsilon}(k, +) \cdot \vec{\varepsilon}(k,-) = 0$$

As discussed in sect. 1.4, the field operator associated to the system of photons is

$$\vec{A}(x) = (2\pi)^{-3/2} \sum_{\lambda = \pm 1} \int \frac{d^3 k}{2\,\omega_k} \vec{\varepsilon}_{k\lambda} \, [a_{k\lambda} \, e^{i\vec{k}.\vec{x}} + a^*_{-k\lambda} \, e^{i\vec{k}.\vec{x}}] \tag{1.25}$$

The hermitean combination of a and a^* is taken in order to have an hermitean field $\vec{A}(x)$, the quantum analogue of a real classical field. As we will see $\vec{A}$ describes the quantum vector potential in the Coulomb gauge ($A_o = 0$, $\mathrm{div}\vec{A} = 0$)

To show that $\vec{A}(x)$ obeys the Maxwell equations one checks that the Hamiltonian

$$H = \int d^3k \; \omega_k \; a^*_{k\lambda} a_{k\lambda}$$

when written in terms of $\vec{A}(x)$, coincides with the Maxwell Hamiltonian

$$H = \frac{1}{8\pi} \int d^3x \left[\left(\frac{\partial \vec{A}}{\partial t} \right)^2 + (\text{curl } \vec{A})^2 \right] = \frac{1}{8\pi} \int d^3x \; (\vec{E}^2 + \vec{H}^2)$$

($\vec{E} \equiv -\partial A/\partial t$, $\vec{H} \equiv \text{curl } \vec{A}$). The time evolution of $\vec{A}$ is clearly governed by the time evolution of a, a^*

$$a_{k\lambda}(t) = e^{-i\omega_k t} \; a_{k\lambda}(0) \quad \text{etc.}$$

The CCR for a, a^*

$$[a_{k\lambda}, a^*_{q\lambda'}] = \delta_{\lambda\lambda'} \; \delta(k - q),$$

$$[a , a] = 0 = [a^*, a^*],$$

determine the commutation relations for $A_i(\vec{x})$, $\dot{A}_j(\vec{y})$ and therefore for $\vec{E}$ and $\vec{H}$

$$[A_i(\vec{x},t), \; A_j(\vec{y},t)] = 0,$$

$$[A_i(\vec{x},t), \; \dot{A}_j(\vec{y},t) = i\left(\delta_{ij} - \frac{\partial_i \partial_j}{\Delta} \right) \delta(\vec{x} - \vec{y}), \tag{1.26}$$

$$[E_i(\vec{x},t), \; H_j(\vec{y},t) = i\varepsilon_{ijk} \partial_k \; \delta(\vec{x} - \vec{y}), \tag{1.26'}$$

(where $\Delta \equiv \sum_i \partial_i^2$ and the following relation has been used

$$\sum_\lambda \varepsilon^{i*}_{k\lambda} \; \varepsilon^j_{k\lambda} = \left(\delta_{ij} - \frac{k_i k_j}{|\vec{k}|^2} \right)$$

For the applications it will be useful to know also the commutator at different times

$$[A_i(\vec{x},t), A_j(\vec{x}',t')] = \left(\delta_{ij} - \frac{\partial_i \partial_j}{\Delta} \right) D(x - x'), \tag{1.27}$$

$$D(x) = \frac{i}{(2\pi)^3} \int \frac{d^3k}{\omega_k} \; e^{i \vec{k} \vec{x} - i\omega_k t} \quad ,$$

D(x) is called the Pauli Jordan function. It is worthwhile to note that the commutator of $A_i(x,t)$, $A_j(y,t')$ does not satisfy the locality (or causality) property i.e. it does not vanish when (x,t) is spacelike with respect to (y,t'). Locality is instead satisfied by the commutators of $\vec{E}$ and $\vec{H}$. The commutation relations (1.26') imply that it is not possible to simultaneously measure $\vec{E}$ and $\vec{H}$ with extreme precision. Bohr and Rosenfeld have deduced the relations (1.26') exclusively by arguments based on Heisenberg uncertainty principle; here we derived them from the corpuscolar properties of the e.m. field.

It is worthwhile to mention that the mean value of $\vec{A}$ and also of $\vec{E}$ and $\vec{H}$ vanish on states with definite photon number

$$< n | \vec{A} | n > = 0 = < n | \vec{E} | n > = < n | \vec{H} | n >$$

whereas the expectation value of the energy $E^2 + H^2$ does not vanish. This means that on those states the classical limit is not obvious. The classical limit is easily obtained by considering the <u>coherent states</u> studied by Glauber(*). For simplicity we first consider the electromagnetic field in a box of volume $V = L^3$ with periodic boundary conditions and we consider just one mode say $\vec{k},\lambda$. The corresponding coherent state is labelled by a complex number z and defined by

$$|z_{k\lambda}> = e^{-\frac{1}{2}|z|^2} \sum_n \frac{z^n}{\sqrt{n!}} | n(\vec{k},\lambda) > \tag{1.28}$$

with $|n(\vec{k},\lambda>$ denoting a state with n photons in the $\vec{k},\lambda$ mode. The interesting

(*) R.J. Glauber, Phys. Rev. Letters <u>10</u>, 84 1963; Phys. Rev. <u>131</u>, 2766 (1963). For a general account about coherent states see J.R. Klauder and E.C.G. Sudarshan, <u>Fundamental of Quantum Optics</u>, W. Benjamin 1968

property is that the expectation value of $\vec{E}$ on $|z_{k\lambda}>$ is a classical electric field of complex amplitude $z_{k\lambda}$ (in the $\vec{k},\lambda$ mode):

$$< z|\ \vec{E}(\vec{x},t)\ | z > = \frac{i}{\sqrt{2}}\ \sqrt{\omega_k}\ \ \vec{\varepsilon}_{k\lambda}\ [\ z\, e^{i(\vec{k}\vec{x} - \omega_k t)} - z^*\ e^{-i(\vec{k}\vec{x} - \omega_k t)}$$

The above equation easily follows from

$$a_{k\lambda}\ |\ z_{k\lambda}\ > \ = \ z_{k\lambda}\ |\ z_{k\lambda}\ >$$

$$< z|\ a\ |z > = z\ , \qquad < z|\ a^*\ |z > \ = z^* \qquad (1.29)$$

The first relation is a characteristic property of coherent states, which can be defined as eigenstates of the destruction operators $a_{k\lambda}$. From the above relations (1.29) it also follows that

$$< z\ |\ a^*\ a\ |\ z > \ = |z|^2 \ = < N_{k\lambda} >$$

$$(\ \Delta N_{k\lambda}\ )^2 = < z\ |\ \ (N_{k\lambda} - < N_{k\lambda} >\ )^2\ |z > \ = \ < N_{k\lambda} > \qquad (1.29')$$

and

$$\Delta N_{k\lambda}\ /\ < N_{k\lambda} > \ = \ < N_{k\lambda} >^{-1/2}$$

The relative uncertainty of the photon number in a coherent state goes to zero as $< N_{k\lambda} > \to \infty$, i.e. in the limit $|z_{k\lambda}| \to \infty$, corresponding to very intense classical electric fields (laser beams).

1.7 Coherent state representation. Infrared catastrophe

The above definition (1.28) of coherent states can also be written as

$$|z_i> = e^{z_i a_i^* - z_i^* a_i} |0> = e^{-\frac{1}{2}|z_i|^2} e^{z_i a_i^*} |0> \equiv U(z_i) |0>$$

($i = k\lambda$) and it can be generalized to infinite degrees of freedom, i.e. to all modes. Let $\{z_i\}$ be a sequence of complex numbers such that

$$\sum |z_i|^2 < \infty$$

then one can introduce a corresponding coherent state by

$$|\{z_i\}> = \exp\left[-\tfrac{1}{2}\sum_i |z_i|^2\right] \sum_{\{n_i\}=0}^{\infty} \prod_{k=1}^{\infty} \frac{z_k^{n_k}}{n_k!} |\{n_k\}>$$

$$= \exp\left[\sum_i (z_i a_i^* - z_i^* a_i)\right] |0> = U(\{z_i\})|0> . \qquad (1.30)$$

One can easily verify that

$$U(\{z_i\})^{-1} a_i U(\{z_i\}) = a_i + z_i \qquad (1.31)$$

i.e. U is a shift operator. Eqs.(1.30) and (1.31) easily imply

$$a_i |\{z_i\}> = z_i |\{z_i\}>$$

The above formulae maintain a meaning also in the infinite volume limit, with $\{z_{k\lambda}\} \to$ a complex function $f(k, \lambda)$.

Given a coherent state $|f>$ (or $|z>$) one can construct a representation of the CCR in the following way. For any polynomial $P(a,a^*)$

$$<f| P(a,a^*)|f> = <0| P(a+f, a^*+f^*) |0> \qquad (1.32)$$

The Hilbert space is then the closure of the set of states of the form

$$\Psi = P(a,a^*) |f> \qquad (1.33)$$

The physical meaning of the so obtained representation is very simple. The above states describe photons in the presence of a radiation field which has a well defined classical limit. As long as $f(k,\lambda) \in L^2$ the representation is unitary equivalent to a Fock representation. The above representation can however be introduced even if $f \notin L^2$, without making reference to the definition (1.30), but by using eq. (1.32) as defining equation for $|f\rangle$ and for the states of the form (1.33). Formally $|f\rangle$ plays a role similar to that of the ground state in the Fock representation, each state being described in terms of elementary modifications of (i.e. as results of application of $P(a,a^*)$ to) the state $|f\rangle$.

Quite generally if

$$A_i = a_i + f_i \tag{1.34}$$

with $\sum_i |f_i|^2 = \infty$, then a representation which is Fock for the operators A_i (as the representation defined by eq. (1.32)) cannot be a Fock representation for the operators a_i, and conversely.

An example which requires the use of non-Fock coherent state representation is provided by the radiation field associated to a (classical) charged particle which changes its velocity. More precisely we consider the limiting situation in which a particle moves with constant velocity $\vec{v}$, is kicked at $t = 0$ and then moves with velocity $\vec{v}\,'$.

The field equation for the (transverse) electromagnetic potential are

$$\Box \vec{A}(x) = \vec{j}(x) \tag{1.35}$$

By eq.(1.25) we have

$$\vec{a}(\vec{k},t) \equiv \sum_{\lambda} \vec{\varepsilon}(\vec{k},\lambda)\, a_{\lambda}(\vec{k},t) = (2\omega)^{-1/2}\,[\,\omega\vec{A}(\vec{k},t) + i\dot{\vec{A}}(\vec{k},t)\,]$$

and therefore eq. (1.35) becomes

$$i\frac{d}{dt}\vec{a}(\vec{k},t) = \omega\,\vec{a}(\vec{k},t) - (2\omega)^{-1/2}\,\vec{j}(\vec{k},t)$$

which can be easily integrated to

$$a_i(\vec{k},t) = e^{-i\omega(t-t_0)}\,a_i(\vec{k},t_0) + i(2\omega)^{-1/2}\,e^{-i\omega t}\int_{t_0}^{t} e^{i\omega t'}\,j_i(\vec{k},t')dt'$$

Chosing the initial time $t_0 \to -\infty$, the above equation becomes

$$a_i(\vec{k},t) = e^{-i\omega t}\,a_i^{in}(\vec{k}) + i(2\omega)^{-1/2}\,e^{-i\omega t}\int_{-\infty}^{t} e^{i\omega t'}\,j_i(\vec{k},t)dt'$$

For the process described above, denoting by θ the Heaviside function, one has

$$\vec{j}(\vec{x},t) = e\,[\theta(t)\vec{v}\,'\,\delta(\vec{x} - \vec{v}\,'t) + \theta(-t)\,\delta(\vec{x} - \vec{v}t)\vec{v}] \tag{1.36}$$

and therefore

$$a_i(\vec{k},t) = e^{-i\omega t}\,a_i(k) = e^{-i\omega t}\,[a_i^{in}(\vec{k}) + f_i(k,t)\,] \tag{1.37}$$

with

$$f_i(\vec{k},t) = \frac{e}{\sqrt{2\omega}}\left[v_i'\,\frac{e^{i(\omega-\vec{k}.\vec{v}\,')t} - 1}{\omega - \vec{k}.\vec{v}\,'} + \frac{v_i}{\omega - \vec{k}.\vec{v}}\right] \tag{1.37'}$$

Since for low $\vec{k}$, $f(k,t) \sim k^{-3/2}$, one has $f \notin L^2$ and therefore, if one chooses a Fock representation for the radiation field in the distant past (i.e. for a^{in}) this representation will not be a Fock representation for the radiation field at the time t. From a physical point of view this means that when undergoing a scattering process a charged particle emits an infinite number of infrared photons (infrared catastrophe). It is obvious from the above equation that, in this simple case, the transition from in-operators to the operators at time t is just provided by a shift operator; insisting on using a Fock representation both for in as well as for operators at time t would lead to the so-called infrared divergences.

1.8. General properties of states of infinite systems

As discussed in the previous sections, the description of a system with infinite degrees of freedom involves i) the choice of the canonical variables, say $\{a, a^*\}$, and the construction of the algebra $\mathcal{A}$ generated by them (algebraic structure), and ii) the identification of the relevant representation π of $\mathcal{A}$, as an algebra of operators in a Hilbert space H, the space of states of the system (dynamical structure). Clearly a representation of $\mathcal{A}$ is fully determined by specifying a Hilbert space H and all the matrix elements $\langle \Psi, A\Psi \rangle$, $\Psi \in H$, $A \in \mathcal{A}$.

A large class of physically interesting representations has the property that the states of the system can be analyzed in terms of the ground(*) state Ψ_o and its elementary excitations, in the sense that the generic state of the given representation can be written as

$$\Psi = P(a, a^*)\, \Psi_o \tag{1.38}$$

(with P a suitable polynomial) or a (strong) limit of states of that form(**). This means that in π the canonical variables are (algebraically) complete in the sense that their algebraic functions suffice for the description of the states of the system. Representations satisfying the above property are therefore totally determined by the set of correlation functions

(*) More generally in terms of a "reference" state.

(**) More precisely, the set of vectors of the form (1.38) form a dense set in the Hilbert space H in which the representation is defined. As mentioned before, to avoid domain problems one should better use as canonical variables the Weyl operators associated to a, a*.

$$< \Psi_o , \quad a_1 \ldots a_n \qquad a^*_{i_1} \ldots a^*_{i_m} \Psi_o > , \tag{1.39}$$

since any matrix element $< \Psi , A\Psi >$, $A \varepsilon \mathcal{A}$ can be reduced to matrix elements of the above form (*). Technically, representations having the above property are said to have a cyclic ground state. Correlation functions of the type (1.39) with canonical variables occurring at different times are usually called Green functions. The knowledge of the Green functions does not only determine the states of the system but also their time evolution, and it is therefore equivalent to the full solution of the dynamical problem.

In most of the following chapters we will focus our attention on physical systems characterized by a translationally invariant ground state:

$$U(\vec{a})\, \Psi_o = \Psi_o , \qquad \forall \vec{a},$$

where $U(\vec{a})$ is the translation operator. This implies that the correlation functions are translationally invariant, i.e. in x-space they are functions of the differences $x_j - x_i$.

An important set of information on the states of infinite systems comes from the realization that in preparing or defining a state, only loca-

(*) In more detail, one can show that given a set of correlation functions (1.39) satisfying the positivity condition: $< A^*A > \geq 0$ for any $A \varepsilon \mathcal{A}$, one may define a vector space D_0, whose elements are labelled by polynomials of the canonical variables, Ψ_P , and a semidefinite product $< \Psi_{P_1}, \Psi_{P_2} > \equiv < P_1^* P_2 >$. D_0 is thus a pre-Hilbert space which after completion and quotient yields the Hilbert space of the representation.

lized measurements (*) are available in our laboratory. Therefore, starting from a given state Ψ_o, one may only produce local excitations. To exploit this important point it is convenient to regard the algebra $\mathcal{A}$ as generated by canonical variables which have some localization properties, like e.g. $\psi(x)$, $\psi^*(x)$ or better the operators

$$\psi(f_\alpha) \equiv \int \psi(x) f_\alpha(x) d^3x \,, \qquad \psi^*(g_\alpha) \tag{1.40}$$

with f_α, g_α localized functions. The algebraic characterization of localization, as it will be needed in the following, is that $\mathcal{A}$ is generated by elements satisfying the following condition (which we will call asymptotic locality):

$$\lim_{|\vec{x}| \to \infty} [\, A_{\vec{x}} \,, B \,] = 0 \tag{1.41}$$

where $A_{\vec{x}}$ denotes the $\vec{x}$-translated of A. The physical meaning of condition (1.41) is very clear: the "measurements" of localized operators do not interfere when the distance of the localization regions goes to infinity. Clearly, condition (1.41) is satisfied by the operators $\psi(f_\alpha)$, $\psi^*(g_\alpha)$, with f_α, g_α localized functions.

For an algebra $\mathcal{A}$ generated by canonical variables which satisfy asymptotic locality we have: any irreducible representation with translationally invariant cyclic state, has a unique translationally invariant state.

(*) R. Haag and D. Kastler, Journ.Math.Phys. 5, 818 (1964).
R. Haag, Nuovo Cimento, 25, 287 (1962).

In fact , if the representation space H contains another invariant state Ψ_o' , which without loss of generality can always be chosen orthogonal to Ψ_o, by the cyclicity of Ψ_o there must be a polynomial P of the localized operators which generate $\mathcal{A}$, such that

$$< \Psi_o', P \Psi_o > \neq 0 \tag{1.42}$$

Now, if $P_{\vec{x}}$ denotes the $\vec{x}$-translated of P, we can construct the ergodic mean

$$\lim_{V \to \infty} \frac{1}{V} \int_V P_{\vec{x}} \, d^3x = P_{av} \tag{1.43}$$

(The limit exists in a weak sense (*)). Clearly for any localized operator A

$$[P_{av} , A] = 0$$

since by eq. (1.41)

$$\lim_{|\vec{x}| \to \infty} [P_{\vec{x}} , A] = 0$$

and the ergodic limit coincides with the ordinary limit when the latter exists. Thus P_{av} commutes with all the elements which generate $\mathcal{A}$ and therefore it commutes with $\mathcal{A}$. In any irreducible representation P_{av} must therefore be a multiple of the identity; this is in contrast with eq.(1.42) since

$$< \Psi_o' , P_{av} \Psi_o > = < \Psi_o' , P \Psi_o >$$

Another interesting property is that the uniqueness of the translationally

---

(*) See e.g. the discussion by R. Haag, in Critical Phenomena, lectures at the Sitges Int.School on Stat.Mech., June 1976, Springer Lect.Notes in Physics, Vol. 54

invariant state is equivalent to the validity of the weak cluster property:

$$\text{ergodic-}\lim_{|\vec{x}|\to\infty} [< \Psi_o , A_{\vec{x}} B \Psi_o > - < A >_o < B >_o] = 0 \qquad (1.44)$$

for any A, B $\in \mathcal{A}$. In fact, by Von Neumann ergodic theorem (*) the ergodic mean of a unitary operator $U(\vec{x})$

$$\lim_{V\to\infty} \frac{1}{V} \int U(\vec{x})\, d^3x = P_{inv}$$

is the projection P_{inv} on the suspace H_o of vectors which are invariant under $U(\vec{x})$. Clearly eq.(1.44) holds if and only if H_o is one dimensional.

The physical meaning of the cluster property is that the correlation between operators localized in very distant regions, factorize; i.e. far separated variables behave independently as far as expectation values are concerned (this factorization is necessary for the construction of the scattering matrix!).

Representations with unique ground state are the equivalent of the pure phases in Statistical Mechanics. This characterization in terms of irreducibility and validity of the (weak) cluster property justifies focusing our attention on them.

One may justify the restriction to representations with a unique translationally invariant ground state also by the following qualitative considerations. If Ψ_{o1}, Ψ_{o2} are two transationally invariant orthogonal pure states, by applying to each of them polynomials of the localized

---

(*) See e.g. M. Reed and B. Simon, Methods of Modern Mathematical Physics vol. I, p. 57, Academic Press 1972

(canonical) variables one obtains two (orthogonal) spaces of states H_1, H_2. From a physical point of view the question is: if states of H_1 can be prepared in a certain laboratory, can one also prepare states of H_2 in the same laboratory? This amount to ask whether states of H_2 can in some way be approximated (*) by states of H_1. Since a state is fully determined by all the possible measurements one can make on it, i.e. by all its expectation values of all the observables, a general criterium of approximation is that given a state Ψ_2 of H_2 there is a sequence of states Ψ_n of H_1 such that

$$(\Psi_n, A\Psi_n) \underset{n\to\infty}{\longrightarrow} (\Psi_2, A\Psi_2) \tag{1.45}$$

for any $A \in \mathcal{A}$. This approximation has a clear physical meaning since it is strictly related to the way a state is prepared in terms of measurements Now, by general qualitative arguments which become particularly convincing or can be made rigorous in some simple model (like spin models of the Ising or Heisenberg type, or the $(\phi^4)_2$ quantum field theory model), in order to get an approximation of the type (1.45) one needs to use states Ψ_n with energy E_n which diverges as $n \to \infty$. Therefore states of H_2 cannot be prepared by starting from states of H_1, if only a finite amount of energy is available. H_1 and H_2 are therefore "disjoint worlds" and only one can be chosen to describe the states of the system which can be prepared in a given laboratory. At the classical level, the appearance of "disjoint worlds" or "disjoint Hilbert

(*) Clearly this approximation cannot be in the strong or even in the weak topology of H_1.

space sectors"stable under time evolution has been proved in general to follow from the non-linearity of the problem for hyperbolic equations(*).

(*) See C. Parenti, F. Strocchi and G. Velo, Comm.Math.Phys. 53, 65 (1977) and the lectures by F. Strocchi in Topics in Functional Analysis 1980-81, Scuola Normale Superiore 1982.

II. INTERACTIONS WITH INFINITE DEGREES OF FREEDOM

2.1 Genuine infinite systems

The description of infinite systems described before, in terms of a ground state and its elementary excitations,becomes particularly useful when the interaction is such that the number of elementary excitations is not a constant of motion. In the time evolution an arbitrarily large number of degrees of freedom may be excited and therefore one has a genuine infinite system with the difficulties and the interesting features of QM_∞.

Ordinary QM is recovered when: i) the Hamiltonian commutes with the number N of elementary excitations, ii) the dynamics selects a representation of the algebra of the canonical variables such that N is a well defined operator. Condition ii) excludes the case of collective effects like condensation, for which N is no longer well defined, and spontaneous breaking of gauge transformations of the first kind (see Part C).

When conditions i) and ii) are satisfied the dynamics is defined in the Fock representations and one can consider the subspaces $H^{(n)} \equiv P^{(n)} H$, with definite number of elementary excitations. The Schroedinger equation for a generic state Ψ can then be written as a system of decoupled independent equations for each component $\Psi^{(n)} \epsilon H^{(n)}$

$$i \frac{d}{dt} \Psi^{(n)} = H_{nn} \Psi^{(n)} , \quad H_{nn} = P^{(n)} H P^{(n)} ,$$

and in each $H^{(n)}$ one has an ordinary QM for n particles. The situation changes when $[N,H] \neq 0$. In this case one has an infinite system of coupled equations

$$i \frac{d}{dt} \Psi^{(n)} = H_{nm} \Psi^{(m)} , \quad H_{nm} = P^{(n)} H P^{(m)} ,$$

i.e. a genuine infinite system. In the following sections we will discuss simple examples in which the number of particles is not a constant of motion.

2.2 Photon emission and absorption by atoms

The use of creation and annihilation operators allows a very simple description of photon emission and absorption processes in atomic transition. To this purpose we consider a physical system consisting of atoms and electromagnetic radiation in a cubic box of volume $V = L^3$, with periodic boundary conditions. In the approximation in which the nuclei are considered as infinitely heavy and the electrons are treated non-relativistically, the atomic system, in the absence of electromagnetic radiation, is described by the following Hamiltonian (see sect. 1.5)

$$H_{atoms} = \int \psi^*(x) \left(\frac{\vec{p}^{\,2}}{2m} + V(x) \right) \psi(x)\, d^3x + e^2 H_{e-e}$$

where V(x) is the Coulomb potential which keeps the electrons bound to the nuclei and $e^2 H_{ee}$ is the electron-electron electrostatic interactions. The interaction with the e.m. radiation is essentially determined by the minimal coupling or gauge invariance ($\vec{p} \to \vec{p} - \frac{e}{c} \vec{A}(x)$) and the total Hamiltonian becomes

$$H = H_o^{rad} + \int \psi^*(x) \left\{ \frac{[\vec{p} - \frac{e}{c}\vec{A}(x)]^2}{2m} + V(x) \right\} \psi(x)\, d^3x + e^2 H_{e-e} \tag{2.1}$$

The eigenstates of

$$H_o \equiv H_o^{rad} + H_{atoms}$$

can be described in the Fock representation and since the Hamiltonian commutes with the number of electrons (and therefore of atoms) we can restrict ourselves to the subspace with a definite number of atoms, for example one atom. The eigenstates of H_o are then of the form $|\ell_1\rangle |n_1, n_2 \ldots\rangle$ where $|\ell_j\rangle$ denotes the state of the atom in the ℓ_j level and $|n_1, n_2 \ldots\rangle$ denotes the state with n_1 photons in the state $|\alpha_1\rangle$ etc. (see sect. 1.6). By treating $\vec{A}(x)$ as a field operator and expanding it in terms of creation and annihilation operators (as in sect. 1.6, eq.(1.25)) one immediately sees that H does not commute with the photon number. The atomic transition

$$\ell_j + n_i \rightarrow \ell_{j'} + n_i - 1$$

with the absorption of a photon in the i-th mode is then governed, to the first order, by the amplitude

$$A_{n_i \rightarrow n_i - 1} \quad \langle \ell_{j'} + n_i - 1 \,|\, H_{int} \,|\, \ell_j + n_i \rangle$$

where

$$H_{int} = \frac{-ei}{mc} \int \psi^*(x) \; \vec{\nabla} \psi(x) \; \vec{A}(x) \; d^3x - \frac{e^2}{2mc^2} \int \psi^*(x) \psi(x) \; \vec{A}^2(x) \; d^3x \tag{2.2}$$

(since the electron-electron interaction Hamiltonian $e^2 H_{e-e}$ gives vanishing contribution to the above matrix element) and only the first term in eq.(2.2) contributes to A. For the evaluation of A one has

$$\langle \ell_j \,|\, \psi^*(x) \; \vec{\nabla}\psi(x) \,|\, \ell_{j'} \rangle = \Psi_j^*(x) \; \vec{\nabla} \; \Psi_{j'}(x)$$

with Ψ_j^* , $\Psi_{j'}$ denoting the atomic wave functions (see sect. 1.5), and

$$<\ldots n_{k_i \lambda_i} - 1 \ldots |\vec{A}(x)| \ldots n_{k_i \lambda_i} \ldots > = \frac{1}{\sqrt{V}} \sqrt{n_{k_i\lambda_i}} \; \vec{\varepsilon}(k_i, \lambda_i) \frac{e^{i\vec{k}_i.\vec{x}}}{\sqrt{2\omega_i}}$$

Hence

$$A_{n_i \to n_i - 1} = - \frac{ie}{mc} \sqrt{n_i} \int \Psi_j^* (x) \vec{\nabla} \Psi_{j'}(x) \vec{\varepsilon}(k_i, \lambda_i) \quad e^{i\vec{k}_i\vec{x}} \frac{1}{\sqrt{2V\omega_i}} \, d^3x$$

i.e. the amplitude is proportional to the square root of the intensity of the present radiation in the i-th mode (see sect. 1.6 eq.(1.29')).More interesting is the emission process

$$\ell_j + n_i \to \ell_{j'} + n_i + 1$$

for which one gets

$$A_{n_i \to n_i + 1} = - \frac{ie}{mc} \sqrt{n_i + 1} \int \Psi_j^*(x) \vec{\nabla} \Psi_{j'}(x) \frac{\vec{\varepsilon}_i e^{-i\vec{k}_i . \vec{x}}}{\sqrt{2V\omega_i}} \, d^3x$$

The transition probability (per unit time) is then proportional to $n_i + 1$. The contribution corresponding to $n_i = 0$ (no e.m. radiation present) is called <u>spontaneous emission</u> , whereas the contribution proportional to n_i is called induced emission. The coefficients n_i, $n_i + 1$ for the absorption and emission of photons by atoms were first deduced by Eistein in 1916 with arguments based on the statistical properties of the e.m. radiation. The (more complete) derivation outlined here as a simple result of the quantization of the e.m. radiation was first discussed by Dirac in 1927.

2.3 Nuclear forces (Yukawa model)

In analogy with the interaction between charged particles mediated by the e.m. field and given by $j_\mu(x)A_\mu(x)$, with $j_\mu(x)$ the e.m. current, in 1935 Yukawa suggested that a similar mechanism could explain the nuclear forces. A scalar meson field $\phi(x)$ was assumed to mediate the nuclear forces(*) through an interaction of the form (non- relativistic approximation)

$$H_{int} = g \int j(x)\,\phi(x) d^3x$$

where $j(x)$ is the nucleon charge density

$$j(x) = \psi^*(x)\psi(x)$$

and $\psi(x)$ is the field operator which describes nucleons. The elementary excitations or particles associated to the field $\phi(x)$ are called mesons and in terms of creation and annihilation operators $\phi(x)$ is written as(**)

$$\phi(x) = (2\pi)^{-3/2} \int \frac{d^3k}{\sqrt{2\omega_k}} \left[a(k)e^{ikx} + a^*(k)e^{-ikx} \right] \qquad (2.3)$$

where $\omega_k^2 \equiv \vec{k}^2 + m^2$, m = meson mass.

To simplify the discussion we do not consider spin effects and will treat $\psi(x)$ as a fermionic field operator satisfying anticommutation relations (see sect. 1.4 eq.(1.19)).In the non-relativistic approximation mentioned

(*) In 1947 the above idea was experimentally proved to be correct with the only difference that the meson field was a pseudoscalar field(pionic field) and the interaction was pseudoscalar ($j(x) \sim \bar{\psi}(x)\,\gamma_5\,\psi(x)$).

(**) To simplify the discussion we consider a neutral field.

above, the model is governed by the following Hamiltonian

$$H = \int \omega_k a^*(k)\, a(k)\, d^3k + \int E(p)\psi^*(p)\psi(p)\, d^3p$$
$$+ \frac{g}{(2\pi)^{3/2}} \int d^3p \frac{d^3k}{\sqrt{2\omega_k}} \psi^*(p+k)\psi(p)\,[\,a(k) + a^*(-k)]$$

with

$$E(p) = \sqrt{p^2 + M^2} \to M + \frac{p^2}{2M}$$

Since the nucleons are treated non-relativistically the model cannot be expected to describe correctly the emission and absorption of high energy mesons and therefore an additional information must be added by hand, by introducing a nucleon form factor F(k). The interaction Hamiltonian then becomes

$$H_{int} = \frac{g}{(2\pi)^{3/2}} \int d^3p \frac{d^3k}{\sqrt{2\omega_k}} F(k)\, \psi^*(p+k)\psi(p)\,[\,a(k) + a^*(-k)] =$$
$$= g \int \psi^*(x)\psi(x)\, \tilde{F}(x-y)\, \phi(y)\, d^3x\, d^3y \qquad (2.4)$$

This amounts to treating the nucleon as an extended object, since the interaction density is not a local operator and it involves an <u>elementary</u> coupling between the meson field $\phi(y)$ and the nucleon density $j(x)$, even when the center of the nucleon is localized in x and the meson is localized in y. The nucleonic extension is described by the Fourier transform of the form factor F(k). The local limit (or the nucleon pointlike limit) is obtained by letting

$$\tilde{F}(x - y) \to \delta(x - y)$$

(i.e. $F(k) \to 1$). From a technical point of view the rôle of the form factor F(k) is that of guaranteeing the convergence of the integrals when needed, so that the expressions we will derive are all mathematically well defined.

The interaction Hamiltonian describes the elementary processes shown in Fig. 2.

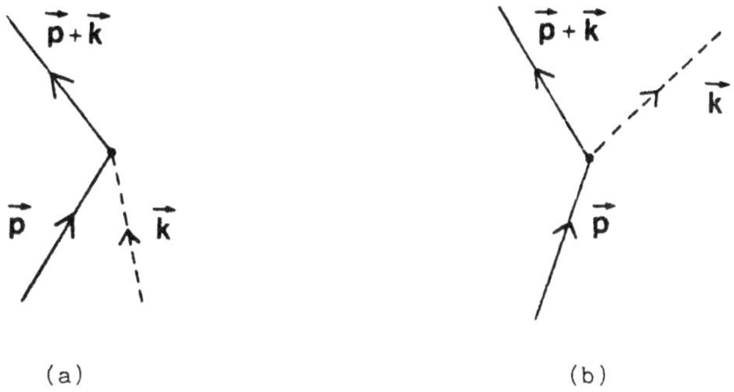

Fig. 2 Meson absorption (a) and emission (b) by a nucleon. The vertex is given by the form factor $gF(k)/\sqrt{2\omega_k}$.

The number of nucleons

$$N = \int \psi^*(x)\psi(x)\, d^3x = \int \psi^*(p)\psi(p)\, d^3p$$

commutes with the Hamiltonian (*) so that we can try to discuss the dynamical problem for each eigenspace of N. It is easy to see that the total momentum

$$\vec{P} = \int d^3p\, \vec{p}\,\psi^*(p)\psi(p) + \int d^3k\, \vec{k}\,a^*(k)\, a(k)$$

is also a constant of motion.

(*) In a fully relativistic theory the meson nucleon interaction would also allow for nucleon-pair creation or annihilation.

We start by looking at the N = 0 sector. A simple computation shows that the ground state $|0\rangle$ of the non-interacting theory (g = 0) is also an eigenstate of H, ($g \neq 0$). It is therefore reasonable to choose a Fock representation for the operators a, a^*, ψ, ψ^* corresponding to the elementary excitations of the non-interacting system. As we will see later this is justified if $m > 0$ and F(k) is a sufficiently decreasing function when $k \to \infty$. In this case, the n (non-interacting) meson states $a^*(k_1)\ldots a^*(k_n)|0\rangle$ are also eigenstates of the total Hamiltonian with eigenvalue $\Sigma\,\omega(k_i)$. Therefore, in the N = 0 sector the dynamics is rather trivial.

The dynamics is more interesting in the $N \neq 0$ sectors, where, as we will see, collective effects appear. To simplify the discussion we will consider the model in the extreme non-relativistic approximation for the nucleon:

$$E(p) \to M, \tag{2.5}$$

i.e. the nucleons are considered so heavy that the momentum dependence of the energy is neglected. The main advantage of this approximation is that the model can be solved exactly and all its interesting features will be under control. We consider the N = 1 sector. The state of one unperturbed nucleon with momentum $\vec{p}$, $\psi^*(\vec{p})\,|0\rangle$ is not an eigenstate of H. The corresponding eigenstate of H, with N = 1 and $\vec{P} = \vec{p}$, is obtained as solution of the eigenvalue equation

$$H\,|\vec{p}\rangle = M'\,|\vec{p}\rangle\ .$$

One obtains a change of the mass

$$M' = M - \frac{g^2}{(2\pi)^3} \int d^3k \frac{|F(k)|^2}{2\omega^2(k)} \equiv M - g^2 \delta M \qquad (2.6)$$

and the following eigenstate

$$|\vec{p}> = e^{iS} \psi^*(p)|0> = \int d^3q \int \frac{d^3x}{(2\pi)^3} e^{-i(\vec{q}-\vec{p})\vec{x}} \psi^*(q) \cdot \qquad (2.7)$$

$$\cdot \exp\left[- \frac{g}{(2\pi)^{3/2}} \int \frac{d^3k}{\sqrt{2\omega_k^3}} F(k) e^{-i\vec{k}.\vec{x}} (a^*(k) - a(-x)) \right]$$

where (*)

$$S = \frac{ig}{(2\pi)^{3/2}} \int d^3p \, d^3k \frac{F(k)}{\sqrt{2\omega^3(k)}} \psi^*(p)\psi(p+k) \left[a^*(k) - a(-k) \right]$$

When analyzed in terms of elementary excitations of the non-interacting system ($g = 0$) the (physical) nucleon state $|\vec{p}>$ is a superposition of a "bare" nucleon and infinite mesons (" <u>meson cloud</u> "):

$$< \vec{p} \, | \frac{1}{\sqrt{n!}} a^*(k_1) \ldots a^*(k_n) \psi^*(q)|0> = \sqrt{Z} \quad \delta^3(\vec{q} + \Sigma \vec{k}_i - \vec{p}) \cdot$$

$$\cdot \frac{(-g)^n}{n!} \prod_{i=1}^{n} \frac{F^*(k_i)}{\sqrt{2(2\pi)^3 \omega^3(k_i)}} ,$$

$$Z \equiv \exp\left[\frac{-g^2}{(2\pi)^3} \int d^3k \frac{|F(k)|^2}{2\omega^3} \right] \qquad (2.8)$$

(*) To derive eq.(2.7) it is useful to note that $e^{-iS}|0> = |0>$ and to expand $e^{iS}\psi^* e^{-iS}$ in powers of S.

The transformation

$$\psi^*(p) \rightarrow e^{iS}\, \psi^*(p)\, e^{-iS} \tag{2.9}$$

is called a "dressing" transformation because it transforms the creation operator of a "bare" nucleon into a creation operator of a physical nucleon. It is interesting to note that infinite degrees of freedom of the non-interacting system contribute to the definition of a physical nucleon state. This is a first example of a collective effect. Such effects are characteristic features of QM_∞, where the interaction Hamiltonian does not only give rise to transient effects like scattering as in ordinary QM, but it also leads to a complete redefinition of the degrees of freedom or elementary excitations of the system, with respect to the $g = 0$ case.

We can now discuss the validity of the Fock representation used so far. First, in the local limit, $F(k) \rightarrow 1$, the integral occurring in the definition of Z is logarithmically divergent for $k \rightarrow \infty$ (ultraviolet divergence), so that $Z \rightarrow 0$. This means that in the local limit the states of the interacting system cannot be analyzed in terms of elementary excitations of the non-interacting system. Equivalently, in the local limit one cannot choose a Fock representation for a, a^*, ψ, ψ^*. The same difficulaty appears also if $F(k)$ decreases sufficiently fast for $k \rightarrow \infty$, but the mesons have zero mass, $(m = 0)$. In this case, the integral occurring in Z is again logarithmically divergent but for $k \rightarrow 0$ (infrared divergence). Again one cannot choose a Fock representation for a, a^*, ψ, ψ^*.

It should also be stressed that in the local limit M' is linearly ultraviolet divergent and therefore the spectrum of the total Hamiltonian in the $N \neq 0$ sectors reduces to the point $-\infty$ unless a suitable subtraction or renormalization procedure is adopted. This means that the model can be extrapolated to very small distances (ultraviolet region) at the price of loosing predictivity about the nucleon mass. The removal of the ultraviolet cutoff, $(F(k) \to F_\Lambda(k) = \theta(|k| - \Lambda),\ \Lambda \to \infty)$ must be supplemented by a prescription about $\lim_{\Lambda \to \infty} M'_\Lambda \equiv M'_\infty$. In order to get a finite value for M'_∞ we have to introduce a cutoff dependence of the "bare" parameter (*) $M : M \to M_\Lambda$, such that $M'_\Lambda = M_\Lambda - g^2 \delta M_\Lambda$ converges to a prescribed finite value M'_∞ as $\Lambda \to \infty$. From a physical point of view the conclusion is that the nucleon mass, a parameter which is very sensitive to the small distance structure of the nucleon cannot be predicted without a more detailed theory of the high energy behaviour. It is however a gratifying feature of the model, common also to all the so-called renormalizable quantum field theories, that the lack of knowledge about a more fundamental theory, which correctly describes the small distance behaviour, can be totally taken care of by specifying a few physical parameters, here the nucleon mass, which can no longer be predicted and become free parameters. Since, in general, the so-called "bare" parameters in the unrenormalized Hamiltonian,(here the bare nucleon mass M),are not measurable quantitites, prescribing the renormalized parameters,(here M'_∞), involves the same amount of freedom as fixing the bare ones and it is clearly more physical. What is lost in this type of theories is the possibility of a purely dynamical explanation of the mass since an infinite "bare" mass counterterm is needed to compensate the divergence of the dynamical

---

(*) Equivalently a cutoff dependent counterterm.

contribution $g^2\delta M_\Lambda$; therefore the mass becomes a completely free parameter.

An interesting result of the model is the explanation of the nucleon-nucleon potential as a result of meson exchange between the two nucleons. The (newtonian) concept of force at a distance is therefore derived from the more fundamental concept of field, which involves contact (or local) interactions.

We start discussing the problem by applying perturbation theory methods. We compute the energy shift of a two-nucleon state to lowest non-trivial order. For simplicity we consider the extreme non-relativistic limit in which the two nucleons are regarded as fixed in the positions $\vec{x}_1$ and $\vec{x}_2$, respectively. The restriction of the Hamiltonian to the subspace characterized by this condition becomes

$$H = \int \omega_k a^*_k a_k d^3k + 2M + g \int [F(x_1 - y) + F(x_2 - y)]\phi(y)d^3y$$

(since $\psi^*(\vec{x})\psi(\vec{x}) |\vec{x}_1, \vec{x}_2> = [\delta(\vec{x} - \vec{x}_1) + \delta(\vec{x} - \vec{x}_2)] |\vec{x}_1, \vec{x}_2>$).

At zero order the energy is 2M and the first order contribution vanishes since $<2| H_{int} | 2> = 0$. The second order gives

$$\Delta E = \sum\!\!\!\!\!\!\int_n \frac{<2| H_{int} |n> <n| H_{int} |2>}{2M - E_n}$$

It is immediate to see that the only intermediate states which contribute are the states with 2 nucleons plus a meson, and the above formula gives

$$\Delta E = - \frac{2g^2}{(2\pi)^3} \int \frac{d^3k}{2\omega_k^2} |F(k)|^2 [1 + e^{i\vec{k}\cdot(\vec{x}_1 - \vec{x}_2)}] \tag{2.10}$$

$$\equiv \Delta E_1 + \Delta E_2$$

The first term ΔE comes from the 1 in the square brackets and it is independent from the relative position, $\vec{x}_2 - \vec{x}_1$, of the two nucleons; it is in fact the dynamical contribution to the mass discussed above. The second term ΔE_2 describes the potential energy of the two nucleons and in the local limit it becomes

$$\Delta E_2(x_1 , x_2) \to - \frac{g^2}{4\pi} \frac{e^{-m|\vec{x}_1 - \vec{x}_2|}}{|\vec{x}_1 - \vec{x}_2|} \tag{2.11}$$

i.e. the so-called Yukawa potential . The range of this potential is $1/m$ and from knowledge of the range of nuclear forces Yukawa predicted the meson mass. With m = pion mass one gets $1/m = 1.4 \times 10^{-13}$ cm. In the limit $m = 0$ one obtains the Coulomb potential.

It is instructive to represent the two contributions diagramatically. ΔE_1 is due to the emission and absorption of a meson by one nucleon, whereas ΔE_2 is due to a meson exchange

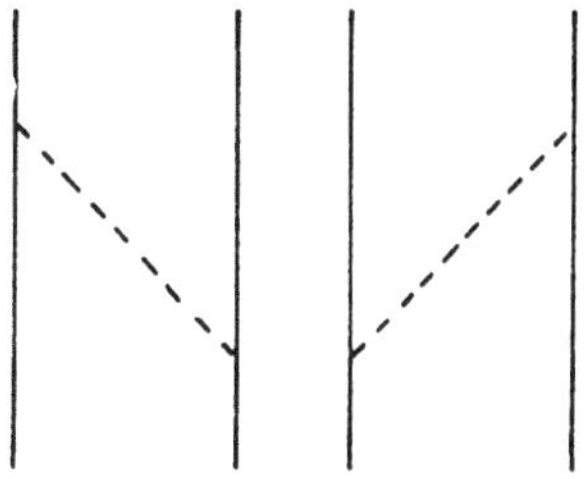

Fig. 3a

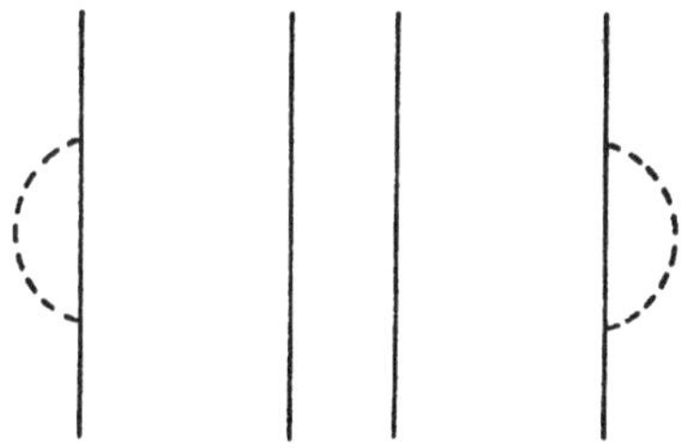

Fig. 3b

The above results are confirmed by the non-perturbative analysis of the model. As shown by Wentzel(*) the second order perturbative calculation of ΔE

(*) G. Wentzel, Quantum theory of Fields, Interscience.

gives the exact result. To this purpose, it suffices to introduce new field opearators A* and Ψ* for mesons and nucleons

$$A^*(k) = e^{iS}\, a^*(k)\, e^{-iS} =$$

$$= a^*(k) + \frac{g}{(2\pi)^{3/2}} \int d^3p\ \psi^*(p+k)\psi(p)\ \frac{F(k)}{\sqrt{2\omega^3}} \qquad (2.12)$$

$$\Psi^*(p) = e^{iS}\ \psi^*(p)\ e^{-iS} = \int d^3q \int \frac{d^3x}{(2\pi)^3}\, e^{-i(\vec{q}-\vec{p}).\vec{x}}\ \psi^*(q)\ .$$
$$.\exp\Big[- \frac{g}{(2\pi)^{3/2}} \int d^3k\ \frac{F(k)}{2\omega_k^3}\ e^{-i\vec{k}.\vec{x}}\ (a^*(k) - a(-k))\Big] \qquad (2.12)$$

in terms of which the Hamiltonian takes a particularly simple form

$$H = \int \omega(k)\ A^*(k)\ A(k)\ d^3k + M' \int \Psi^*(p)\Psi(p)\ d^3p +$$
$$+ \int d^3x\ d^3y\ \Psi^*(x)\Psi^*(y)\ V(x-y)\ \Psi(y)\ \Psi(x)\ , \qquad (2.13)$$

where

$$V(x-y) = -\frac{g^2}{(2\pi)^3} \int \frac{d^3k}{2\omega^2}\ |F(k)|^2\ e^{i\vec{k}.(\vec{x}-\vec{y})} \qquad (2.14)$$

The advantage of this new form of the Hamiltonian is that the collective effects have been explicitely exhibited in the bilinear part of the Hamiltonian ($M \to M'$). There is no residual meson-nucleon interaction; the original interaction has been replaced by an effective nucleon-nucleon interaction with a potential which reduces to the Yukawa potential, in the local or pointlike limit. The disappearance of the meson-nucleon interaction is due to the extreme non-relativistic approximation ($E(p) \to M$) made in the definition of the model.

The collective effects leading to a redefinition of the basic parameters of the theory (here the nucleon mass) are also called persistent effects, in contrast with the transient effects typical of a scattering process. It is a characteristic feature of QM_∞ that the interaction Hamiltonian is not only responsible for transient or scattering effects, as in ordinary QM, but it also leads to a redefinition of the degrees of freedom and of the basic parameters of the theory.

The above eqs.(2.13) (2.14) clearly show which representation must be chosen in order to have a meaningful dynamics: it is the Fock representation for the field operators A and Ψ corresponding to the elementary excitations of the interacting system ($g \neq 0$). In the local limit and/or in the $m \to 0$ limit, such excitations cannot be described in terms of the elementary excitations of the non-interacting system ($g = 0$).

The Fock representation for A and Ψ is unitarily equivalent to the Fock representation for a and ψ for $m \neq 0$ and F(k) sufficiently decreasing (the unitary operator which intertwines between the two representations is e^{iS}) but it is not unitarily equivalent for $m = 0$ and/or $F(k) \to 1$.

In the latter cases the definition of A*(k) and Ψ*(k) by the right-hand side of eqs.(2.12) (2.12')still maintains a meaning, but there is no unitary opearator (exp iS becomes meaningless!) which intertwines between the new field operators and the old ones.

2.4 Electron-phonon interactions in metals and in polar crystals

We start by considering the electron-phonon interaction in the case in which the phonons are associated to displacements of charged ions, like in conducting media (acoustical mode). In the approximation in which the ions are considered as very heavy the electrostatic interaction between electrons and ions is described by the following Hamiltonian density

$$\sum_i \rho(x)\, U\,(x - q_i),$$

where $\rho(x)$ is the electron density

$$\rho(x) = \psi^*(x)\,\psi\,(x)$$

and $U(x - q_i)$ is the electrostatic potential due to the i-th ion, at position q_i. For small ion displacements, U can be expanded around the equilibrium position $\vec{q}_i^{\,o} = \vec{\ell}$

$$U(x - q_i) = u(x - q_i^o) + \vec{\nabla}_i U \cdot \delta\vec{q}_i + \ldots$$

The first term gives rise to an interaction term which is independent of the lattice vibrations, whereas the second term gives rise to an interaction Hamiltonian of the form

$$H_{int} = e \sum_i \int \psi^*(x)\,\psi\,(x)\; \vec{\nabla}\, U(x - q_i^o)\cdot \delta\vec{q}_i \; d^3x$$

or, in momentum representation ($M \equiv$ ion mass, $\rho \equiv MN/V$)

$$H_{int} = \frac{e}{(2\pi)^{3/2}} \sum_i \int d^3p\, d^3k\;\; \psi^*(p)\,\psi\,(p-k)\; \frac{i\,\vec{k}\cdot\vec{\varepsilon}\,(k,\lambda)\;\tilde{U}(k)}{\sqrt{2\omega(k)\;\rho}} \;\cdot$$

$$\cdot\; [\, a(k,\lambda)\, e^{i\vec{k}\cdot\vec{q}_i^{\,o}} + a^*(-k,\lambda)\, e^{-i\vec{k}\cdot\vec{q}_i^{\,o}}\,]\;, \qquad (2.15)$$

where the ion displacements $\delta\vec{q}_i$ have been expanded in normal modes, as in sect. 1.3, (<u>acoustical waves</u>).Clearly only longitudinal waves can interact

$$\vec{k} \cdot \vec{\epsilon}\,(k,\lambda) = |\vec{k}| = k$$

and the interaction is very similar to that discussed in the Yukawa model, with a form factor now given by

$$F(k) = \frac{k}{\sqrt{\rho\omega(k)}}\,\tilde{U}(k)$$

One might think that $\tilde{U}(k)$ should be the Fourier transform of the Coulomb potential. Actually, the effective electron-ion potential is not a pure Coulomb potential: it is so only at small distances, whereas al large separations screening effects occur and a good approximation for low k ($k \ll k_F$ = = Fermi momentum) is given by

$$\tilde{U}(k) = \frac{4\pi e}{\vec{k}^2 + \mu^2}\,, \quad \frac{4\pi}{\mu^2} = \frac{\pi^2 a_0}{k_F}$$

with a_o = atomic Bohr. μ^{-1} is therefore of the order of the lattice spacing(*). Thus for low k, since we are dealing with acoustical modes,

$$F(k) \sim C\sqrt{k}$$

The Debye frequency, related to the inverse lattice spacing, provides the ultraviolet cut-off for the integral in the Hamiltonian.

The lattice vibrations in a polar crystal are dominantly of the

---

(*) For a more detailed treatment see A.L. Fetter and J.D. Walecka, <u>Quantum Theory of Many-Particle Systems</u>, Mc. Graw-Hill Book Co. 1971, Sect. 45.

optical type, corresponding to a molecular motion in which the center of mass is at rest and the relative coordinate of the two atoms is responsible for the mode (See sect. 1.3). This gives rise to an optically active dipole $\vec{d}(x) \sim e(q_\ell - q_{\ell+1})$. The electrostatic potential $A_o(x)$ associated to the dipole $\vec{d}(\vec{x})$ is given by

$$\text{grad } A_o(x) = 4\pi \vec{d}(x)$$

or

$$\Delta A_o(x) = 4\pi \text{ div } \vec{d}(x)$$

Hence

$$A_o(k) = 4\pi \vec{k}\cdot\vec{d}(k)/\vec{k}^2$$

and the interaction Hamiltonian

$$H_{int} = e \int d^3x \, \psi^*(x)\, \psi(x)\, A_o(x)$$

becomes

$$H_{int} = \frac{4\pi e^2}{(2\pi)^{3/2}} \int d^3k \, d^3p \;\; \psi^*(p+k)\, \psi(p) \; \frac{\vec{k}\cdot\vec{\varepsilon}\,(k,\lambda)}{\vec{k}^2 \sqrt{2\rho\,\omega(k)}}$$

$$[\; a(k,\lambda) + a^*(-k,\lambda) \;] \tag{2.16}$$

Since for low k, $\omega(k) \sim \omega_o$ independent of k for an optical mode, the Hamiltonian (2.16) is again similar to that of the Yukawa model, but with a form factor

$$F(k) \simeq \frac{k}{k^2} \frac{1}{\sqrt{2\rho\,\omega(o)}} \sim \frac{C}{\sqrt{2}} \; \frac{1}{k}$$

2.5 Polaron model. Mass and coupling constant renormalization

As mentioned before, the Hamiltonian for an electron in a polar crystal (polaron model), eq. (2.16), is very similar to that of the Yukawa model discussed in Sect. 2.3. As before, the number of electrons

$$N_e = \int \psi^*(p)\,\psi(p)\,d^3p$$

and the total momentum

$$\vec{P} = \int d^3k\;\vec{k}\;a^*(k)a(k) + \int d^3p\;\vec{p}\;\psi^*(p)\psi(p)$$

are constants of motion and therefore the eigenstates of H can be classified in terms of the eigenvalues of N_e and $\vec{P}$. However, here we cannot consider the electron in the extreme non-relativistic approximation as for the Yukawa model and the free electron energy cannot be regarded as independent from the momentum:

$$E_o(p) = \frac{p^2}{2m} \quad , \quad m = \text{free electron mass} \tag{2.17}$$

The dependence of E(p) from $\vec{p}$ does no longer allow the model to be exactly soluble and we will treat it perturbatively. As in the Yukawa model, the eigenstate of H corresponding to $N_e = 1$ and $\vec{P} = \vec{q}$, now called polaron, will involve infinite non-interacting phonons, when analyzed in terms of the elementary excitations of the non-interacting system. To the first perturbative order the polaron state is given by

$$|\vec{q}> = |\vec{q},0> + \sum_n \frac{|n,0><0,n|\,H_{int}\,|\vec{q},0>}{E_o(q) - E_n^o(q)} \tag{2.18}$$

where $|q,0>$ and $|n,0>$ are eigenstates of the free Hamiltonian and $E_0(q)$, $E_n^0(q)$ are the corresponding energies. Due to the form of H_{int} only the states with one electron of momentum $\vec{q} - \vec{k}$ plus a phonon of momentum $\vec{k}$, $|\vec{q} - \vec{k}, \vec{k}, 0>$, can contribute to the sum in (2.18) (states with one electron and n phonons will contribute to the n-th perturbative order).

The polaron energy E(q) can be compute by perturbation theory. Since $<0,\vec{q}|\ H_{int}|\vec{q},\ 0> = 0$ the first non-trivial contribution is given by the second order term

$$E(q) = E_0(q) + \int d^3k \frac{|<q-k,k|H_{int}|q>|^2}{E_0(q) - E_0(q-k) - \omega(k)}$$

$$= E_0(q) - \frac{2me^2}{(2\pi)^3} \int d^3k \frac{C^2/2k^2}{\vec{k}^2 - 2kq\cos\theta + 2m\omega(k)}$$

i.e.

$$\Delta E = E(q) - E_0(q) = \frac{m\,e^2C^2}{8\pi^3} 2\pi \int_{-1}^{1} d\cos\theta \int_0^\infty \frac{dk}{k^2 - 2kq\cos\theta + 2m\omega(k)}$$

By approximating $\omega(k) = \omega_0$, the integral can be computed analytically. We will compute it for small q by expanding the denominator in powers of q. Putting

$$x \equiv k/\sqrt{2m\omega_0}\quad,\quad a = q/\sqrt{2m\omega_0}$$

we have

$$\int_{-1}^{1} d\cos\theta\,(2m\omega_0)^{-1/2} \int_0^\infty dx \frac{1}{1+x^2}\left(1 + \frac{2xa\cos\theta}{1+x^2} + \frac{4x^2a^2\cos^2\theta}{(1+x^2)^2} + \ldots\right)$$

$$= (2m\omega_0)^{-1/2} \int_0^\infty dx \frac{1}{1+x^2}\left(2 + \frac{8}{3}a^2\frac{x^2}{(1+x^2)^2} + \ldots\right)$$

and

$$E(q) - E_0(q) = - \alpha (\omega_0 + \frac{q^2}{12m} + \dots), \quad \alpha = \frac{e^2}{4\pi} \sqrt{\frac{m}{2\omega_0}} \; C^2$$

The term $-\alpha\omega_0$ independent of q, is actually the ground state energy shift due to the electron-phonon interaction. The term $-\alpha q^2/12m$ implies a change in the mechanical properties of the electron as a result of the interaction with the phonon system:

$$E(q) = - \alpha\omega_0 + \frac{q^2}{2m^*} \quad , \; m^* = m/(1 - \frac{\alpha}{6})$$

i.e. the polaron mass is m^* and differs from the free electron mass (mass renormalization).

As for the Yukawa model we can compute the interaction potential between two electrons at positions $\vec{x}_1$, $\vec{x}_2$ inside a polar crystal. To second perturbative order we get

$$V(\vec{x}_1 - \vec{x}_2) = - \frac{e^2}{8\pi} \frac{C^2}{|\vec{x}_1 - \vec{x}_2|} = - \frac{e^2}{4\pi\varepsilon} \frac{1}{|\vec{x}_1 - \vec{x}_2|}$$

with $\varepsilon \equiv 2/C^2$. The potential is still of the Coulomb type, but with a change of the coupling constant

$$e^2 \to e^2/\varepsilon \quad ,$$

(coupling constant renormalization). ε has the meaning of a dielectric constant, which arises here as a collective effect. By measurements of dielectric constant one can determine the constant C and compare its determination in terms of microscopic parameters. Given C and ω_0 one can determine α and m^*.

III. RELATIVISTIC PARTICLE SYSTEMS

3.1 Relativistic field operators

The framework discussed in Chap.I is easily generalized to describe relativistic particle systems. In this case the elementary excitations or particles may be labelled by the momentum $\vec{k}$ and the energy $\omega(\vec{k})$, satisfying the relativistic spectrum condition

$$\omega(\vec{k}) = \sqrt{\vec{k}^2 + m^2} \tag{3.1}$$

(m = the particle mass). The creation and annihilation operators $a(\vec{k})$, $a^*(\vec{k})$ are usually defined in such a way that they obey Lorentz covariant commutation relations. For example, for spinless particles

$$[\ a(\vec{k}),\ a^*(\vec{k}')\ ] = 2k_o\ \delta(\vec{k} - \vec{k}') \tag{3.2}$$

$k_o = \omega(\vec{k})$, and the field operators are defined by

$$\phi(\vec{x},t) \equiv \phi(x) = \frac{1}{(2\pi)^{3/2}} \int \frac{d^3k}{2k_o} [\, a(k)e^{-ikx} + a^*(k)\, e^{-ikx}\,]$$

($kx \equiv \vec{k} \cdot \vec{x} - k_o t$). The Hilbert space for such relativistic system carries a representation of the Poincaré group by unitary operators $U(a,\Lambda)$, such that

$$U(a,\Lambda)\ \phi(x)\ U(a,\Lambda)^{-1} = \phi(\Lambda x + a) \tag{3.3}$$

The Fock representation and related properties carry through in this case as discussed in Chap. I.

3.2 Causality or locality

The relativistic invariance leads in this case to further constraints with respect to the non-relativistic case. The most important one is the causality or locality constraint. One can easily check that the canonical commutation relations (3.2) lead to

$$[\ \phi(x,t),\quad \phi(\vec{y},t')\ ] = 0\ , \tag{3.4}$$

if $(\vec{x},t)$ is spacelike with respect to $(\vec{y},t')$, i.e. field operators at spacelike separated points commute. For fermionic fields eq.(3.4) is replaced by the anticommutation relation

$$[\ \psi(\vec{x},t),\ \psi^*(\vec{y},t')\ ]_+ = 0 \tag{3.5}$$

if $(\vec{x},t)$ is spacelike with respect to $(\vec{y},t')$. The property (3.4), (3.5) is called locality or microscopic causality and it is supposed to have a more fundamental status than the canonical commutation relations (3.2) (or the analogous CAR's). The reason is that, as we will discuss in the following, the relativity group is responsible for a more singular behaviour of the field operators than in the non-relativistic case and the canonical commutation relations loose their meaning in the presence of interactions.

Theorem (Wightman)(*) Let $\phi(x)$ be a hermitean scalar field (**) satisfying the covariance equation (3.3) and let Ψ_o be the unique state invariant

(*) A.S. Wightman, Annales de l'Inst. H. Poincaré, I, 403 (1964).

(**)The theorem extends also to non hermitean fields and also to fields describing particles with spin.

under $U(a, \Lambda)$

$$U(a, \Lambda)\,\Psi_o = \Psi_o \tag{3.6}$$

Then $\phi(x)$ cannot be an operator defined at each point x unless it is the trivial operator $\phi(x)\,\Psi_o = c\,\Psi_o$, c a constant.

Proof. We consider the two-point function

$$<\Psi_o,\ \phi(x)\,\phi(y)\,\Psi_o> \;=\; F(x,y)$$

By eq.(3.3) and (3.6) it is a function of the difference $x - y$ and if $\phi(x)$ is an operator defined at each point, F is a continuous function of $x - y$ since

$$\phi(x)\,\Psi_o = U(x,0)\,\phi(0)\,\Psi_o$$

and $U(x)$ is a (strongly) continuous unitary operator. Furthermore for each sequence of complex numbers α_i

$$\Sigma\,\bar{\alpha}_i\,F(x_i - y_i)\,\alpha_j = ||\,\Sigma\,\alpha_j\phi(x_j)\,\Psi_o||^2 \geq 0$$

i.e. F is a function of positive type(*) and by Bochner theorem(*) its Fourier transform is a finite positive measure

$$F(x) = \int e^{ipx}\,d\mu(p) \tag{3.7}$$

with

$$\int |d\mu(p)| < \infty \tag{3.8}$$

(*) See L. Schwartz, Théorie des Distributions, Tome II, Ch. VII, § 9, Hermann 1951.

The transformation law (3.3) of ϕ under the Lorentz group implies

$$\mu(\Lambda p) = \mu(p)$$

i.e. μ is a Lorentz invariant measure.According to Gårding analysis(*) μ has the following form

$$d\mu(p) = b\,\delta(p) + c\, d\rho(p^2)\,\frac{d^3p}{\sqrt{p^2+\vec{p}^2}}$$

with b, c constants, and eq.(3.8) requires c = 0.This implies that only the ground state can occur as intermediate state in the two point function, that $\langle\Psi_0, \phi(x)\,\Psi_0\rangle = \sqrt{b}$ and $(\phi(x) - \sqrt{b}\,)\,\Psi_0 = 0$; thus $\phi(x)$ is a trivial operator.

A similar result holds (**) if the covariance eq.(3.3) is replaced by

i) the covariance under space time translations

$$U(a)\,\phi(x)U(a)^{-1} = \phi(x+a)$$

with the generator P^μ satisfying the relativistic spectrum condition $(p_0 \geq 0,\ p^\mu p_\mu \geq 0)$

ii) locality of the fields

$$[\,\phi(x),\ \phi(y)\,] = 0 \qquad \text{if } (x-y)^2 < 0$$

The above results show that strictly speaking one cannot introduce the algebra of the field operators $\phi(x)$, but only the algebra of the regularized or smeared operators

$$\phi(f) = \int \phi(x)f(x)\,d^4x \tag{3.9}$$

(*) L. Gårding and J.L. Lions, Nuovo Cimento Supp. 14, 45 (1959).
(**) A.S. Wightman, Annales dé l'Inst. H. Poincaré, I, 403 (1964).

with f(x) a sufficiently regular test function (typically $f \in C^{\infty}$ and of compact support). These problems become more acute in the presence of interactions, when one can no longer require the validity of the CCR's (or the CAR's). The singular behaviour of the fields is amplified by the interaction and one cannot in general consider them at sharp time; only the smearing in space and time guarantees that one gets well defined (possibly) unbounded operators. Otherwise one would end up with CCR's of the form

$$[\ \phi(\vec{x},t),\ \partial_o\ \phi(\vec{y},t) \ = i\ Z_3^{-1}\ \delta(\vec{x}-\vec{y}) \tag{3.10}$$

with Z_3 a constant (the so-called wave function renormalization constant) which is infinite when computed in perturbation theory, unless there is no interaction.

For relativistic quantum field theory one must therefore give up the algebraic constraint of the CCR's (or CAR's). Its place is taken by the causality or locality constraint

$$[\ \phi(f),\ \phi(g)\ ] = 0 \tag{3.11}$$

if support of f is spacelike with respect to supp of g (and similarly for fermion fields). In conclusion, for relativistic systems, the algebra of the canonical variables must be replaced by the field algebra $\mathcal{A}$, generated by polynomials of the smeared fields $\phi(f)$, with the algebraic constraint given by locality, eq.(3.11), (local field algebra).

3.3 Ground state and local excitations. Cluster property and uniqueness of the ground state

As for non-relativistic systems, a representation of the local field algebra is specified by the so-called ground state Ψ_o (vacuum state), which is expected to be cyclic with respect to the polynomials of the fields $\phi(f)$. This means that the states of the system can be described in terms of the ground state and its local excitations. This is equivalent to say that a representation is totally specified by the so-called correlation functions (Wightman functions)

$$< \Psi_o , \phi(f_i) \ldots \phi(f_n) \Psi_o > \qquad (3.12)$$

The uniqueness of the translationally invariant state is a necessary property for the irreducibility of the representation. The connection with the validity of the cluster property becomes sharper than in the non-relativistic case.

Theorem (Araki-Hepp-Ruelle)(*) In any representation of the field algebra such that

1) there is a unitary representation of the space-time translations with respect to which the basic fields transform covariantly

$$U(a) \phi(x)U(a)^{-1} = \phi(x+a)$$

2) the generators of the space time translations satisfy the relativistic

(*) H. Araki, K. Hepp and D. Ruelle, Helv.Acta Phys. 35, 164 (1962).

spectrum condition ($P_0 \geq 0$, $P^\mu P_\mu \geq 0$)

3) The basic fields satisfy locality

$$[\ \phi(x),\ \phi(y)\ ] = 0 \qquad \text{for } (x - y)^2 < 0,$$

the cluster property (*)

$$\lim_{\lambda \to \infty} [<\ \phi(x_i)\ldots \phi(x_j)\ \phi(y_i + \lambda a)\ldots \phi(y_k + \lambda a) > - < \phi(x_i)\ \ldots\ \phi(x_j)> \ < \phi(y_i)\ldots \phi(y_k) >] \tag{3.13}$$

(a a spacelike vector) is equivalent to the uniqueness of the translationally invariant state.

3.4 General properties of local quantum field theory

The general properties of the states of a relativistic system (quantum field theory) which we have outlined above and justified on physical grounds can be spelled out more precisely (Wightman axioms). A physically relevant representation of the field algebra is required to have the following properties

1) The Hilbert space H contains a unitary representation of the Poincaré group: $\{ a, \Lambda\} \to U(a, \Lambda)$
2) The vacuum state Ψ_0 is the unique translationally invariant state
3) The spectrum of the generators of the space-time translations satisfies the relativistic spectrum condition ($P_0 \geq 0$, $P^2 \geq 0$)

(*) The convergence is in the sense of distributions of $\mathcal{S}'$

4) The fields transform covariantly under the Poincaré group.(For example for a scalar field eq.(3.4) holds).

5) The fields $\phi(f)$ satisfy the locality condition

$$[\ \phi(f),\ \phi(g)\ ] = 0$$

if supp f is spacelike with respect to supp g.

A more systematic and complete discussion of the above general properties does not fall into the scope of these notes. We refer to the excellent presentation by R.F. Streater and A.S. Wightman, PCT, Spin and Statistics and All That, Benjamin 1964. Here we want to mention that the above properties can be equivalently state in terms of corresponding properties of the correlation functions

$$< \phi(x_1)\ldots \phi(x_n) > \equiv \mathcal{W}(x_1,\ldots x_n) = W(\xi_1, \ldots \xi_{n-1})$$

$\xi_n \equiv x_n - x_{n-1}$, namely

A) The Wightman functions are Poincaré covariant. For a scalar field

$$W(\Lambda\xi + a) = W(\xi)$$

B) The support of the Fourier transform of $W(\xi)$ is contained in the forward cone

$$\tilde{W}(q) = 0 \qquad \text{if} \qquad q \notin \bar{V}^+$$

C) The cluster property (3.13) holds

D) The locality condition holds

$$\mathcal{W}(\ldots x_i\, x_{i+1},\ldots) = \mathcal{W}(\ldots x_{i+1}, x_i, \ldots)$$

if $(x_i - x_{i+1})^2 < 0$

E) The positivity condition holds

$$W(f^* \times f) = \sum_n \int W_n(x_i \ldots x_n) \sum_{\ell+k=n} f^*(x_1)\ldots f^*(x_\ell)$$

$$f(x_{\ell+1})\ldots f(x_n)\, dx_1 \ldots dx_n \geq 0$$

It is the content of Wightman reconstruction theorem that a set of correlation functions satisfying A) - E) uniquely determine (up to isomorphisms) a representation of the field algebra satisfying 1) - 5).

PART B – COLLECTIVE EFFECTS. CONDENSATION

I. ELECTRON GAS

1.1 Free electron gas. Bogoliubov transformation

We consider a system of N electrons in a cubic box of side L, with periodic boundary conditions. For each electron the wave vector $k = (k_1, k_2, k_3)$ can take the discrete values $k_i = (2\pi/L)n_i$, $n_i = 0,1,\dots,$ so that the one particle states are described by lattice points (n_1, n_2, n_3) and by a spin variable $s_z = \pm \frac{1}{2}$. The ground state is then characterized by the electrons occupying the lowest energy states, compatibly with Pauli principle. Each electron state occupies a "cell" in k space of volume $\frac{1}{2}(2\pi/L)^3$ (the factor $\frac{1}{2}$ is due to the spin multiplicity) and therefore the ground state is characterized by a density

$$\rho(k) = 2(L/2\pi)^3/L^3 \qquad \text{for } k \le k_F$$

$$\rho(k) = 0 \qquad \text{for } k > k_F$$

where k_F is the maximum value of k which is occupied. k_F is called the radius of the Fermi sphere and it is determined by the electron density

$$n = \frac{N}{L^3} = \int_{k \le k_F} \rho(k) d^3k = \frac{2}{(2\pi)^3} \frac{4}{3}\pi\, k_F^3 = \frac{1}{3\pi^2} k_F^3 .$$

The energy E_F of an electron on the Fermi surface is

$$E_F \equiv \frac{\hbar^2}{2m} k_F^2 = \frac{\hbar^2}{2m} (3\pi^2 n)^{2/3} \tag{1.1}$$

It is useful to introduce the <u>characteristic radius r_o</u> associated to the volume per electron

$$\frac{V}{N} = \frac{1}{n} \equiv \frac{4}{3}\pi r_o^3 \tag{1.2}$$

or the dimensionless parameter

$$r_s \equiv r_o/a_o \quad , \qquad a_o \equiv \text{Bohr radius.}$$

The energy per electron in the ground state is

$$\frac{E}{N} = \frac{2}{N}\left(\frac{L}{2\pi}\right)^3 \int d^3k \, \frac{\hbar k^2}{2m} = \frac{3}{5} E_F = \frac{2.21 \text{Ryd}}{r_s^2} \, . \tag{1.3}$$

The quantities k_F, E_F, r_s depend only on the density n and are therefore stable under the thermodynamical limit ($V \to \infty$).

The ground state does not satisfy the Fock condition with respect to the electron creation and annihilation operators :

$$a(k,s)\,\Psi_o \neq 0 \qquad \text{for } k < k_F$$

This implies that a, a* do not describe the elementary excitation of the free electron gas. They are better described in terms of the modifications of the ground state, namely by specifying the number of electrons with $k > k_F$ and the number of unoccupied states inside the Fermi sphere, called "holes". In this picture the excitation of an electron from the ground state ($k < k_F$) to a state with $k > k_F$ is described by a creation of a hole of energy $|\varepsilon(k)| \equiv |E(k) - E_F|$ and a creation of an excited state above the Fermi surface with energy $E(k) - E_F \equiv \varepsilon(k)$. This suggests to make a change of canonical variables

$$\begin{aligned} c(k,s) &= a(k,s), && \text{for } k > k_F, \\ d(k,s) &= a^*(-k,-s), && \text{for } k < k_F. \end{aligned} \tag{1.4}$$

(For simplicity, in the following, we will sometimes omit the spin variable s). The choice of $a^*(-k,-s)$ instead of $a^*(k,s)$ in eq.(1.4) guarantees that the transformation (1.4) commutes with translations and rotations in the infinite volume limit.

It is very simple to see that the ground state Ψ_o satisfies the Fock condition for the operators c and d and the Hamiltonian becomes

$$H = \Sigma E(k)\, a^*(k)a(k) = \sum_{k<k_F} \varepsilon(k) + \sum_{k<k_F} |\varepsilon(k)|\, d^*_k d_k + \sum_{k>k_F} \varepsilon(k)\, c^*_k c_k + E_F N_{op}$$

where $\varepsilon(k) \equiv E(k) - E_F$ and N_{op} is the total particle number operator. By choosing E_F as zero energy point for each elementary excitation, $H \to H - E_F N_{op}$, one gets the new Hamiltonian

$$H_o = \sum_{k<k_F} \varepsilon(k) + \Sigma |\varepsilon(k)|\, A^*_k A_k \tag{1.5}$$

with $A_k = c_k$ for $k > k_F$, $A_k = d_k$ for $k < k_F$.

The tranformation (1.4) is a special case of the so-called <u>Valatin-Bogoliubov tranformation</u>

$$\begin{aligned} A_k &= u_k a_k - v_k a^*_{-k} \equiv \alpha(a_k) \\ A^*_k &= u^*_k a^*_k - v^*_k a_{-k} \equiv \alpha(a^*_k) \end{aligned} \tag{1.6}$$

For bosons, the canonicity condition implies

$$\begin{aligned} |u_k|^2 - |v_k|^2 &= 1 \\ u_k v_{-k} - v_k u_{-k} &= 0 \end{aligned} \tag{1.7}$$

i.e.

$$u_k = e^{i\alpha_k} \cosh\theta_k, \qquad v_k = e^{i\beta_k} \sinh\theta_k,$$

$$e^{i(\beta_k - \alpha_k)} \mathrm{tgh}\, \theta_k = e^{i(\beta_{-k} - \alpha_{-k})} \mathrm{tgh}\, \theta_{-k}$$

Without loss of generality we can consider the case $\alpha_k = 0$ and obtain the solution

$$\theta_k = \theta_{-k}, \tag{1.8}$$

$$\beta_k = \beta_{-k}$$

Similarly for fermions we get

$$|u_k|^2 + |v_k|^2 = 1 \tag{1.9}$$

$$u_k v_{-k} + v_k u_{-k} = 0$$

with solution

$$u_k = e^{i\alpha_k} \cos\theta_k, \qquad v_k = e^{i\beta_k} \sin\theta_k$$

$$e^{i(\beta_k - \alpha_k)} \mathrm{tg}\theta_k = - e^{i(\beta_{-k} - \alpha_{-k})} \mathrm{tg}\, \theta_{-k}$$

Chosing $\alpha_k = 0$ we have the solution

$$\theta_k = - \theta_{-k}$$

$$\beta_k = - \beta_{-k} \tag{1.10}$$

The Fock representation for A_k, A_k^* is unitarily equivalent to the Fock representation for a_k, a_k^* only if the following condition holds

$$\sum_k |v_k|^2 < \infty \tag{1.11}$$

Condition (1.11) can be easily derived by requiring that the number operator corresponding to the new variables

$$N = \sum_k A_k^* A_k$$

is well defined in the Fock representation for a_k, a_k^*, i.e. on the Fock ground state Ψ_o for a_k, a_k^*. For finite volume V, the above condition (1.11) is clearly satisfied by the transformation (1.4). Quite generally, in the infinite volume limit,

$$\sum_k |v_k|^2 \longrightarrow V \int d^3k |v(k)|^2 \rightarrow \infty$$

and therefore condition (1.11) can never be satisfied if $\theta(k) \neq 0$. In the infinite volume limit the states with simple interpretation in terms of elementary excitations associated to A_k, A_k^* cannot be analysed in terms of Fock states for a_k, a_k^*. The inequivalence of the two representations in the infinite volume limit can be directly seen by noting that, in a (irreducible) Fock representation, the no-particle state Ψ_o is the unique translationally invariant state and therefore, if there is a unitary operator exp i Q which induces the transformation (1.6), exp i Q Ψ_o must be proportional to Ψ_o; this is possible only if Q = 0.

Before closing this section it is worthwhile to remark that for the free electron gas the energy per unit volume

$$\frac{E}{V} = \frac{3}{5}\frac{\hbar^2}{2m}(3\pi^2 n)^{2/3} n$$

does not have a minimum for non-zero density. Therefore in the thermodynamical limit ($V \rightarrow \infty$) the lowest energy state is the state with zero density ($N/V \rightarrow 0$). To get a ground state with non-zero density one must fix N/V = n in advance and perform the limit $V \rightarrow \infty$ by keeping n fixed. In this way, however the density n becomes an external parameter, which is put in by hand and there is no possibility of predicting it. States with $n \neq 0$ would actually appear unstable towards a lowering of the density. As we will see in the following sections the interaction will stabilize the electron gas, with the energy per unit volume attaining its minimum at a finite density (fermion condensation).

1.2 Interacting electron gas. Hartree-Fock approximation

The model for a quantum theory of metals is a system of electrons interacting with one another and with the lattice ions via the Coulomb potential. Since the ion mass is much larger than the electron mass, the ions will be regarded as fixed. In terms of electron creation and annihilation operators the hamiltonian is

$$H = H_o + H_{int},$$
$$H_o = \int d^3x\ \psi^*(x)\, h(x)\psi(x),\ h(x) = -\frac{\hbar^2\Delta}{2m} + U(x), \tag{1.12}$$
$$H_{int} = \tfrac{1}{2}\int d^3x\, d^3y\ \psi^*(x)\psi^*(y)\ V(x-y)\ \psi(y)\psi(x),$$

here m is the electron mass, U(x) is the Coulomb potential due to the ions and V(x-y) is the electron-electron Coulomb potential. If $f_i(x)$ denotes the single-particle wave function for the i-th state and a^*_i the corresponding creation operator, so that

$$\psi(x) = \sum_i f_i(x) a_i ,$$

The Hamiltonian (1.12) may be written as

$$H = \sum_{ij} h_{ij}\ a^*_i\ a_j + \tfrac{1}{2} \sum_{ijk\ell} a^*_i a^*_j a_k a_\ell\ v_{ijk\ell} , \tag{1.13}$$

with

$$h_{ij} = \int d^3x\ f^*_i(x)\ h(x)\ f_j(x),$$
$$v_{ijk\ell} = \int d^3x\, d^3y\ f^*_i(x)\ f^*_j(y)\ V(x-y)\ f_k(y)\ f_\ell(x) . \tag{1.13'}$$

The assumption at the basis of the Hartree-Fock approximation is that the eigenstates of H and in particular <u>the ground state can be described in terms of single particle wave functions.</u> This means that the ground state for an N electron system is approximately of the form $|N; n_1, n_2 \ldots\rangle$, with <u>suitable</u>

wave functions $f_i(x)$ characterizing the single particle states. The dynamical problem is then reduced to the determination of the $f_i(x)$; for the ground state they are determined by the condition of minimizing the expectation value of H.

We have

$$< n_1,n_2 \dots |H_0| \, n_1,n_2 \dots > \; = \sum_i n_i h_{ii} \, ,$$

$$< H_{int} > = < n_1,n_2 \dots |H_{int}| n_1,n_2 \dots > =$$

$$= \tfrac{1}{2} \sum_{ijk} < a_i^* a_j^* \, a_k a_\ell > v_{ijk\ell}$$

$$= \tfrac{1}{2} \sum_{ij} < a_i^* a_j^* a_j a_i > (v_{ijji} - v_{ijij})$$

since for diagonal matrix elements only $j = k$, $i = \ell$ or $i = k$, $j = \ell$ can contribute and by the CAR's $a_i^* a_j^* a_i a_j = - a_i^* a_j^* a_j a_i$. Moreover, since $a_i^2 = 0$, only the terms with $i \neq j$ survive and, for $i \neq j$, $a_i^* a_j^* a_j a_i = N_i N_j$. Hence

$$< H > = \sum_i n_i h_{ij} + \tfrac{1}{2} \sum_{ij} n_i n_j (v_{ijji} - v_{ijij}) \tag{1.14}$$

The above formula leads to a very simple and instructive physical interpretation. By spelling out the spin variables $f_i(x) = f_i(x,\sigma)$ the term v_{ijji} is easily recognized as the classical Coulomb interaction:

$$v_{ijji} = \int d^3x \, d^3y \, f_i^*(x,\sigma) \, f_j^*(y,\sigma')V(x-y)f_j(y,\sigma')f_i(x,\sigma),$$

and

$$\sum_{ij} n_i n_j v_{ijji} = \frac{e^2}{2} \int d^3x \, d^3y \, \frac{< \rho(x)> \, < \rho(y)>}{|x-y|} \, , \tag{1.15}$$

where

$$<\rho(x)> \; \equiv \sum_\sigma < \psi_\sigma(x)\psi_\sigma(x) > \; = \sum_i n_i |f_i(x,\sigma)|^2 .$$

The other term, called <u>exchange interaction term</u>, does not have a classical analog and it appears as a consequence of the antisymmetrization of the wave functions (Pauli principle):

$$\tfrac{1}{2}\sum n_i n_j v_{ijij} = -\tfrac{1}{2}\sum_{ij} n_i n_j \int d^3x\, d^3y \sum_{\sigma} f_i^*(x,\sigma) f_j(x,\sigma) V(x-y) \sum_{\sigma'} f_j^*(y,\sigma') f_i(y,\sigma') =$$

$$= -\frac{e^2}{2} \int d^3x\, d^3y \sum_{\sigma\sigma'} \frac{|\langle \rho_{\sigma\sigma'}(x,y)\rangle|^2}{|x-y|}, \qquad (1.16)$$

where

$$\langle \rho_{\sigma\sigma'}(x,y)\rangle \equiv \langle \psi_\sigma(x)\ \ \psi_{\sigma'}(y)\rangle = \sum_i n_i f_i^*(x,\sigma) f_i(y,\sigma')$$

The term (1.16) describes an effective spin dependent Hamiltonian as it can be easily seen in the limit of no spin-orbit coupling, $f_i(x,\sigma) \to g_i(x)\chi_i(\sigma)$. In fact,

$$\sum_{\sigma} \chi_i(\sigma)^\dagger\ \chi_j(\sigma)$$

is non zero only if the electrons in the i-th and j-th states have parallel spins and therefore the exchange terms is essentially equivalent to an effective spin Hamiltonian of the form (Heisenberg Hamiltonian)

$$H_{spin} = \tfrac{1}{2} \sum_{ij} J_{ij}(1 + \vec{\sigma}_i \cdot \vec{\sigma}_j) \qquad (1.17)$$

Even if the Coulomb interaction is described by a spin independent potential, the Pauli principle gives rise to a coupling between the i-th and j-th spins. It is useful to remark that the exchange terms is negative and therefore spin allignement is favoured in the ground state. The above considerations are in fact at the basis of Heisenberg theory of ferromagnetism.

For the determination of the single particle wave functions for the ground state, it is convenient to incorporate the orthogonality condition into the variational equation by means of a Lagrange multiplier λ_{ij}:

The variation with respect to f_j yields the Hartree-Fock equation

$$-\frac{\hbar^2}{2m}\Delta f_j(x,\sigma) - U f_j(x,\sigma) + e^2 \int d^3y \frac{<\rho(y)>}{|x-y|} f_j(x,\sigma) - \qquad (1.18)$$
$$- e^2 \sum_{\sigma'} \int d^3y \frac{<\rho_{\sigma'\sigma}(y,x)>}{|x-y|} f_j(y,\sigma') = \sum_k \lambda_{jk} f_k(x,\sigma)$$

Since, in the expression (1.14) for $<H>$, the terms corresponding to $i = j$ in the direct Coulomb interaction are exactly cancelled by the $i = j$ terms in the exchange interaction, one may write the H-F equation (1.18) with the reduced densities

$$<\tilde{\rho}(y)>_j \equiv \sum_{\sigma, i\neq j} n_i f_i^*(y,\sigma) f_i(y,\sigma)$$

$$<\tilde{\rho}_{\sigma'\sigma}(y,x)>_j \equiv \sum_{i\neq j} n_i f_i^*(y,\sigma') f_i(x,\sigma)$$

which do not contain the charge densities of the j-th electron.

The operator on the left hand side of eq.(1.18) is hermitian and therefore by a change of basis one may always reduce to the case in which the matrix λ_{ij} is diagonal

$$\lambda_{ij} = \varepsilon_i \delta_{ij}$$

For large N, the parameter ε_j has the meaning of the energy of the j-th electron in the presence of the other electrons and of the fixed ions. In fact, for large N, the wave functions $f_i(x)$, $i \neq j$, are not expected to change appreciably if the j-th electron is removed and the corresponding change in energy is

$$< n_1,\ldots,n_j,\ldots|H|n_1,\ldots n_j,\ldots > - < n_1,\ldots n_{j-1}\ldots|H|n_1,\ldots n_{j-1},\ldots > =$$
$$\sum_{\sigma} \int d^3x\, f_j^*(x,\sigma). \quad [\text{ left-hand side of eq. (1.18)}] = \varepsilon_j$$

With the above interpretation of ε_j the H-F equation can be read as a Schrödinger equation for the wave function of the j-th electron, in the presence of the potential $U(x)$ due to the ions, of the Coulomb potential due to the charge distribution $< \tilde{\rho}(x) >$ and of the potential due to the exchange charge distribution of the other electrons.

The H-F equations (1.18) form a set of N coupled non-linear equations and a possible approach to their solution is a self-consistent procedure. One starts with a properly chosen set of functions f_i, one computes the corresponding $\tilde{\rho}$'s and then determines the new $f_i(x, \sigma)$ by the H-F equations. The procedure is then iterated self-consistently.

The H-F equations can be directly obtained from the non-linear field equations corresponding to the Hamiltonian (1.12),

$$i \frac{\partial}{\partial t} \psi(x,t) = - [H, \psi(x)] = (- \frac{\hbar^2}{2m} \Delta + U(x))\psi(x) +$$

$$+ \int \psi^*(y) V(x-y) \psi(y) \psi(x) d^3y$$

by a mean field approximation which replaces $\psi^*(y)\psi(y)$ in the second term and $\psi^*(y)\psi(x)$ in the second term on the left-hand side by their expectation values. The above substitution gives rise to the direct and the exchange term, respectively.

The non-perturbative control of the Hartree-Fock approximation (and also of other approximations made in the following sections) is a (basic) open problem of many-body theory and the fundamental question remains whether the results obtained by those uncontrolled approximations reflect the real behaviour of the theory(*).

(*) For a discussion of these problems and a general account on the general investigations on these questions we refer to: W. Thirring, Quantum Mechanics of Large Systems, Springer Verlag 1983.

1.3 Ground state energy. Fermion condensation

A simple solution of the H-F equations can be easily found if the ion density is approximated by a uniform density

$$\rho_{\text{ions}} = e\,\frac{N}{V}$$

(jellium model). In this case the plane waves (*)

$$f_i(x,\sigma) = \frac{1}{\sqrt{V}}\, e^{ik_i x}\, \chi_i(\sigma)$$

are solutions of eq.(1.18). Even if it is not clear that the above functions minimize $\langle H \rangle$, we will compute the ground state energy under this approximation.

Since for plane waves the charge density is uniform,

$$\langle \rho(x) \rangle = N/V$$

the direct Coulomb interaction (the third term in eq.(1.18)) is exactly cancelled by ion electrostatic potential. It remains to compute the exchange term. For plane waves we have

$$\langle \rho_{\sigma'\sigma}(y,x) \rangle = \frac{1}{V} \sum_i e^{i\vec{k}_i(\vec{x}-\vec{y})}\, \chi_i(\sigma')\chi_i(\sigma)$$

and

$$\sum_{\sigma'} \chi_i^\dagger(\sigma')\chi_i(\sigma)\chi_j(\sigma') = \chi_j(\sigma)$$

so that the exchange term becomes

(*) As usual, for the system in a cubic box of volume V we impose periodic boundary conditions.

$$- \frac{e^2}{V} \sum_i \int d^3y \, \frac{e^{i(\vec{k}_j - \vec{k}_i)\vec{y}}}{|\vec{y}-\vec{x}|} \, e^{i\vec{k}_i\vec{x}} \, \chi_j(\sigma) =$$

$$= - \frac{e^2}{V} \sum_{k_i < k_F} \frac{4\pi}{|\vec{k}_j - \vec{k}_i|^2} \, e^{i(\vec{k}_j - \vec{k}_i)\vec{x}} \, e^{i\vec{k}_i\vec{x}} \, \chi_j(\sigma)$$

$$= \left(\frac{-e^2}{V} \sum_{k_i < k_F} \frac{4\pi}{|\vec{k}_j - \vec{k}_i|^2} \right) e^{i\vec{k}_j\vec{x}} \, \chi_j(\sigma)$$

By approximating the sum over k_i by an integral we get

$$\varepsilon_i(k) = \frac{\hbar^2 k^2}{2m} - \int_{k<k_F} \frac{d^3q}{(2\pi)^3} \frac{4\pi e^2}{|\vec{k} - \vec{q}|} = \tag{1.19}$$

$$= \frac{\hbar^2 k^2}{2m} - \frac{2e^2}{\pi} k_F \, F(k/k_F)$$

where

$$F(x) \equiv \tfrac{1}{2} + \frac{1 - x^2}{4x} \ln \frac{|1 + x|}{|1 - x|}$$

To obtain the total energy we have to sum eq.(1.19) over $k < k_F$, by taking into account the factor 2 in the free part due to spin degeneracy, and the factor ½ in the interaction term to avoid counting the contribution of each pair of electrons, twice. The result is

$$E = 2 \sum_{k<k_F} \frac{\hbar^2 k^2}{2m} - \frac{e^2 k_F}{\pi} \sum_{k<k_F} \left\{ 1 + \frac{k_F^2 - k^2}{2 k k_F} \ln \frac{|k_F + k|}{|k_F - k|} \right\} \tag{1.20}$$

By approximating the sums over k by integrals we get

$$\frac{E}{N} = \frac{3}{5} E_F - \frac{3}{4} e^2 \frac{k_F}{\pi} = \frac{e^2}{2a_o} \left[\frac{2.21}{r_s^2} - \frac{0.916}{r_s} \right] \tag{1.21}$$

where $e^2/2a_o$ = 1 Ryd = 13.6 eV.

It is worthwhile to remark that the second term in eq.(1.21) is due to the exchange potential and corresponds to the interaction of electrons with parallel spin. The important result is that, in the free case the energy per electron does not have a minimum for finite non-zero density, (it goes to zero only when $r_s \to \infty$ i.e. in the limit of zero density), whereas in the interacting case the exchange interaction leads to a minimum for $r_s = 4.83$ with E/N = - 1.29 eV (fermion condensation). For a typical metal like sodium the experimental data give $r_s = 3.96$ and E/N = -1.13 eV, so that the agreement is rather good, especially in view of the approximations made.

1.4. Validity of the Hartree-Fock approximation: high density plasma

The calculation of the previous section relies on assumptions (the Hartree-Fock approximation, the use of plane waves, etc.) which are not obvious. To justify the above procedure one may try to have control on some perturbative expansion which yields the result (1.20) to lowest order. This is not a trivial task for several reasons. First of all, even if the electromagnetic coupling constant may be regarded as small, the interaction Hamiltonian cannot be considered as a small perturbation. As a matter of fact, the infrared singularity of the Coulomb potential cannot be treate perturbatively(*). Collective non-perturbative effects play a relevant role in the physical properties of the electron gas and their treatment requires particular care.

In this section we will discuss the expansion around the high density limit and its connection with the previous result. For the treatment of the infrared singularity of the Coulomb potential we will discuss an infrared regularization and its relation with the thermodynamical limit. In this way we will see in a concrete case how the thermodynamical limit ($V \to \infty$, $N \to \infty$, $N/V = n$ fixed) turns out to be useful for the study of many body systems(**): not only it is suitable for the analysis of intensive quantities but it often leads to a substantial simplification of the dynamical problem. We shall see another example in the BCS model for superconductivity.

(*) D. Bohm and D. Pines, Phys. Rev. 92, 609; 626 (1953)
A.L. Fetter and J.D. Walecka, Quantum Theory of Many Particle Systems, Mc Graw-Hill, New York 1971.

(**) A.L. Fetter and J.D. Walecka, loc. cit.

The first difficulty is that the different terms of the Hamiltonian (1.12) are not well defined because of the infrared divergences associated to the Coulomb potential. For example the ion-electron interaction term can be easily computed by treating the ions as a uniform charge distribution and the electron as plane waves. One gets

$$H_{i-e} \sim -e^2 \left(\frac{N}{V}\right)^2 V\, 4\pi \quad V^{2/3}$$

so that the contribution to the "energy per particle" H_{i-e}/V is divergent as $V \to \infty$. To keep track of the infrared divergences in a correct way it is convenient to introduce an infrared cutoff by replacing the Coulomb potential by a Yukawa potential $e^{-\mu r}/r$ and let $\mu \to 0$ only after the thermodynamical limit $(V \to \infty)$. This is what is prescribed by physical considerations, since to get a reasonable thermodynamical limit the volume size $L = V^{1/3}$ must be much larger than any other length in the problem and therefore the interplay between the infrared cutoff and the volume size must be constrained by

$$\mu^{-1} << L. \tag{1.22}$$

To keep the invariance under translations we impose periodic boundary conditions. One easily obtains

$$\begin{aligned} H_{i-e} &= -e^2 \sum_{k,\sigma} \frac{N}{V} \int d^3x\, d^3x' \,\frac{e^{-\mu|\vec{x}-\vec{x}'|}}{|\vec{x}-\vec{x}'|} = \\ &= -e^2 \left(\frac{N}{V}\right)^2 \int d^3x \quad d^3z \,\frac{e^{-\mu|\vec{z}|}}{|\vec{z}|} \simeq -e^2 \left(\frac{N}{V}\right)^2 \frac{4\pi}{\mu^2} V \end{aligned} \tag{1.23}$$

where the integral over z has been approximated by an integration over an infinite volume. (The change of variables $\vec{x}' \to \vec{z} = \vec{x} - \vec{x}'$ has lead to the above integral over z because of translational invariance).

As a consequence of eq.(1.22), the infrared divergence $\mu^{-2} \to \infty$, as $\mu \to 0$ can show up only if $V \to \infty$ (in such a way that $\mu^{-2} << V^{2/3}$).

Similarly for the ion electrostatic energy H_i one gets

$$H_i = \frac{1}{2} e^2 \left(\frac{N}{V}\right)^2 \int d^3x \quad d^3z \quad \frac{e^{-\mu|\vec{z}|}}{|\vec{z}|} \simeq \frac{1}{2} e^2 \left(\frac{N}{V}\right)^2 \frac{4\pi}{\mu^2} V \tag{1.24}$$

Finally for the term describing the electron-electron interaction one obtains

$$\begin{aligned} H_{e-e} = {} & \frac{e^2}{2V} \sum_{q \neq 0} \sum_{\substack{k,p \\ \lambda_1 \lambda_2}} \frac{4\pi}{q^2 + \mu^2} \, a^{\dagger}_{\lambda_1}(k+q) \, a^{\dagger}_{\lambda_2}(p-q) \, a_{\lambda_2}(p) \, a_{\lambda_1}(k) \\ & + \frac{e^2}{2V} \sum_{k,p,\lambda_1,\lambda_2} \frac{4\pi}{\mu^2} \, a^{\dagger}_{\lambda_1}(k) \, a^{\dagger}_{\lambda_2}(p) \, a_{\lambda_2}(p) \, a_{\lambda_1}(k) \end{aligned} \tag{1.25}$$

where in the discrete sum over q the term $\vec{q} = 0$ has been isolated. Since

$$\sum_{k,p,\lambda_1,\lambda_2} a^{\dagger}_{\lambda_1}(k) \, a^{\dagger}_{\lambda_2}(p) \, a_{\lambda_2}(p) \, a_{\lambda_1}(k) = N^2_{op} - N_{op}$$

and we are interested in computing the energy per particle (E/N) for the ground state, only the term N^2_{op} contributes in the thermodynamical limit of the second term of eq.(1.25); it yields

$$\frac{1}{2} e \left(\frac{N}{V}\right)^2 \frac{4\pi}{\mu^2} V \tag{1.26}$$

The term proportional to N_{op} would give

$$- \frac{e^2}{2} \frac{N}{V} \frac{4\pi}{\mu^2}$$

and therefore a vanishing contribution to the energy per electron (for finite V we can work with states with a definite value of N).

The three terms (1.23) (1.24) (1.25) add up to zero and therefore the total Hamiltonian becomes

$$H = \sum_{p,\lambda} \frac{\hbar^2 p^2}{2m}\, a^{\dagger}_{\lambda}(p)\, a(p) + \text{first term on the r.h.s. of eq. (1.25)}$$

$$= \frac{e^2}{a_o}\, r_s^2\, [\tfrac{1}{2} \sum \bar{k}^2\, a^{\dagger}_{\lambda}(\bar{k})\, a_{\lambda}(\bar{k}) + \qquad (1.27)$$

$$\frac{r_s}{2\bar{V}} \sum \frac{4\pi}{\bar{q}^2+\mu^2}\, a^{\dagger}_{\lambda_1}(\bar{k}+\bar{q})\, a^{\dagger}_{\lambda_2}(\bar{p}-\bar{q})\, a_{\lambda_2}(\bar{p})\, a_{\lambda_1}(\bar{k})$$

where in the second line all the lengths have been measured in units of $r_o \equiv$ size of the volume per electron (eq. (1.2)), (e.g. $\bar{k} \equiv k_o k$, $\bar{V} = r_o^3 V$ etc.), and $r_s = r_o/a_o$, a_o = Bohr radius = $\hbar^2/me^2$.

Thus, with respect to the kinetic part, the interaction appears multiplied by a constant which goes to zero in the high density limit ($r_o \to 0$ or $r_s \to 0$). We have thus obtained an interesting starting point for setting up an expansion around the high density limit. The very simple calculations outlined above show that a careful treatment of the infra-red singularities and of the infinite volume limit lead to a cancellation of the infra-red divergences(*) at the lowest order of a perturbative expansion around $r_s = 0$. The higher orders of such expansion require further analysis. It is not too difficult to see that the second order calculation yields a ground state energy per unit volume which is logarithmically divergent in the infra-red region. In fact.

(*) In terms of diagrams (see e.g. Part A, Sect. 2.3), the infra-red divergence of the sum of diagrams

is cancelled by the "counter terms" describing the interaction with the background of positive ions

$$\frac{E^{2nd}}{V} \simeq \frac{2}{V}\,\frac{1}{(2V)^2}\sum_{k,k',q}\frac{1}{q^4}\,\frac{<(1-n_{k-q})(1-n_{k'+q})n_k n_{k'}>}{-\vec{q}.(\vec{q}+\vec{k}'-\vec{k})}$$

$$\simeq \int d^3q\,d^3k\,d^3k'\;\frac{<\text{ as above }>}{q^4(-\vec{q}).(\vec{q}+\vec{k}'-\vec{k})}$$

Since the term $<\ >$ is non-vanishing only if $k < k_F$, $k' < k_F$, $|\vec{k}-\vec{q}| > k_F$, $|\vec{k}'+\vec{q}| > k_F$, for given $\vec{q}$, $\vec{k}$ and $\vec{k}'$ must be in a layer of thickness $|\vec{q}|$ inside the Fermi surface. Thus for small q the integration of the term $<\ >$ over k and k' lead to a contribution of order q and the integral over q is logarithmically divergent ($\sim \int q^{-1}dq$).

More generally one can see that infra-red divergences occur to all orders of the expansion. A finite result can then be obtained either by ad hoc prescriptions or by a resummation of the leading divergent terms(*) or by taking into account, from the start, that by non-perturbative collective effects the Coulomb potential gets "screened" at large distances(**). The discussion of such procedures falls outside the scope of the present notes and we refer the reader to the corresponding literature.

It is worthwhile to mention that the ground state energy so obtained is not an analytic function of r_s

$$\frac{E_o}{N} = \frac{e^2}{2a_o}\left(\frac{2.21}{r_s^2} - \frac{0.916}{r_s} + 0.062 \ln r_s + 0.096 + O(r_s \ln r_s)\right)$$

(*) M. Gell-Mann and K. Brueckner, Phys. Rev. 106, 364 (1957).

(**) D. Bohm and D. Pines, loc. cit.

II. SUPERFLUIDITY

2.1 Bose-Einstein condensation

The occurrence of a condensation phenomenon, as seen in the previous chapter, leads to a representation of the algebra $\mathcal{A}$ of canonical variables, which is not equivalent to the Fock representation and therefore one is facing the problem of labelling the physically relevant representations. If we start with a representation π_V of the algebra $\mathcal{A}_V$ generated by $\psi(f)$, $\psi^*(g)$ in a finite volume V, the infinite volume limit will in general yield a reducible representation. A reduction into irreducible components can be obtained by finding the operators Q which commute with $\mathcal{A}$ and which therefore reduce to multiples of the identity in any irreducible representation of $\mathcal{A}$. An approach to the identification of such operators Q has been discussed by Haag in connection with the BCS model(*).

To this purpose one considers quasi local operators of the form

$$Q_V = \frac{1}{V} \int d^3x \; dy \; dz \; f(x,y,z)\psi^*(y_1)\ldots\psi^*(y_k)\psi(z_1) \ldots\psi(z_n) \tag{2.1}$$

where y,z are multivariables $y = (\vec{y}_1,\vec{y}_2\ldots\vec{y}_k)$, $z = (\vec{z}_1, \vec{z}_2, \ldots \vec{z}_n)$ and f(x,y,z) is a function which is non-zero only if y and z are in a neighbourhood

(*) R. Haag, Nuovo Cimento 25, 287 (1962).

of x [*]. One then considers the weak limits [**]

$$\underset{V\to\infty}{\text{w-lim}}\; Q_V \equiv Q \tag{2.2}$$

(Haag charges). The important property of the Haag charges is that they commute with $\mathcal{A}$. For example

$$\lim_{V\to\infty} [Q_V, \psi(h)] = \lim_{V\to\infty} \frac{1}{V} \int d^3x'\, d^{3k}y\, d^{3n}z \;\; f(x',y,z)\, n(x) d^3x$$

$$\sum_j \delta(x-y_j)\, \psi^* \ldots \widehat{\psi^*(y_j)} \ldots \psi^*(y_k)\psi(z_1)\ldots \psi(z_n) = 0$$

where the symbol $\frown$ means that the variable $\psi^*(y_i)$ has to be omitted. The Haag charges can therefore be used to label irreducible representations of $\mathcal{A}$.

A simple example is obtained by choosing $f(x,y,z) = \delta^3(\vec{x} - \vec{z})$:

$$Q_o = \underset{V\to\infty}{\text{w-lim}} \int_V d^3x\, \psi(x) = \lim_{V\to\infty} \frac{a_o}{\sqrt{V}}, \tag{2.3}$$

where

$$a_k = \frac{1}{\sqrt{V}} \int_V d^3x\, e^{ikx}\, \psi(x).$$

Thus, if $N_o \equiv a_o^* a_o$ denotes the number of particles with vanishing wave vector k, we get

$$Q_o^* Q_o = \lim_{V\to\infty} \frac{N_o}{V} = n_o,$$

and since Q_o is a c-number

$$Q_o = \sqrt{n_o}\, e^{i\theta} . \tag{2.4}$$

(*) More generally f(x,y,z) is required to decrease rapidly when y or z are far from x: for example $f \in \mathcal{S}$

(**) For a more detailed discussion see R. Haag, in Critical Phenomena, Springer Lect. Notes in Physics, Vol. 54, 1976.

The physically relevant representations of $\mathcal{A}$ or equivalently their ground states are then labelled by the parameters n_o, θ : $|n_o, \theta\rangle$. By translational invariance one easily gets

$$\langle n_o, \theta \,|\psi(x)\,|\, n_o, \theta\rangle = \lim_{V\to\infty} \frac{1}{V} \int d^3x \langle n_o, \theta|\psi(x)|n_o, \theta\rangle = \sqrt{n_o}\; e^{i\theta}$$

and therefore the representations with $n_o \neq 0$ exhibit an order parameter, which is not invariant under the gauge transformation

$$\psi(x) \to e^{i\alpha}\,\psi(x) \qquad \alpha \in \mathbb{R} \tag{2.5}$$

Clearly, ground states labelled by different values of the parameter n_o, define inequivalent representations of $\mathcal{A}$ (Q_0 is a c-number in each irreducible representation and it cannot be mapped into a different c-number by a unitary transformation).

The subalgebra $\mathcal{A}_0 \subset \mathcal{A}$ generated by monomials containing an equal number of ψ's and ψ^*'s is pointwise invariant under the transformation (2.5), (<u>gauge invariant algebra</u>). Since the observables of the theory are elements of $\mathcal{A}_0$, their correlation functions do not depend on the gauge parameter θ. Thus, ground states labelled by the same n_o and by different values of θ define inequivalent representations of $\mathcal{A}$, but equivalent representations of $\mathcal{A}_0$. We will return to these features later, in connection with spontaneous symmetry breaking and gauge quantum field theories.

Finally, we comment on the determination of the ground state for a system of free bosons in the infinite volume limit: $V\to\infty$, $\langle N\rangle/V = n$ fixed. We consider the energy per unit volume h, formally defined by

$$h = \lim_{\to \infty} \frac{H_V}{V} = \sum_k \omega_k \tilde{n}_k \ , \qquad \tilde{n}_k = \frac{N_k}{V}$$

The ground state is characterized by the property of minimizing the expectation value of h, compatibly with the constraint

$$\frac{\langle N \rangle}{V} = \sum_k \langle \tilde{n}_k \rangle = n$$

This conditional minimization problem can be treated by using a Lagrange multiplier and by considering the operator

$$h' = h - \mu \left(\sum_{k \neq 0} \tilde{n}_k + \tilde{n}_o - n \right),$$

where μ has the meaning of a chemical potential. The minimum of $\langle h' \rangle$ as a function of $\langle \tilde{n}_k \rangle$, $\langle \tilde{n}_o \rangle$ and μ is obtained for $\mu = 0$, $\langle \tilde{n}_k \rangle = 0$, $\langle \tilde{n}_o \rangle = n$, i.e. all the particles are in the $k = 0$ state, (total Bose-Einstein condensation). The elementary excitations have energy $\omega(k)$ and for non-relativistic particles $\omega(k) = k^2/2m$.

2.2 Superfluidity

A non-trivial example of Bose-Einstein condensation is offered by the theory of superfluidity, which can also be used to illustrate the general ideas discussed before.

The phenomenon of superfluidity of the liquid Helium is related to the Bose character of the helium atoms and it can be detected only at very low temperatures, where other physical systems are in the solid phase. Schematically, superfluidity can be reduced to the fact that a body of mass M moving with velocity v, lower than a critical velocity v_c inside liquid helium is not slowed down by viscosity at low enough temperatures. At a microscopic level a slowing down by viscosity can be explained in terms of creation of elementary excitations in the fluid, at the expense of energy and momentum by the body of mass M. Thus if $\omega(k)$ is the energy of an excitation of momentum k, the conservation of energy and momentum gives

$$\tfrac{1}{2} Mv^2 = \tfrac{1}{2} Mv'^2 + \omega(k),$$
$$M\vec{v}' = M\vec{v} - \vec{k}$$

By squaring the second equation and by combining it with the first, one gets

$$M^2v'^2 = M^2v'^2 + 2M\omega(k) - 2M\,\vec{v}.\vec{k},$$

i.e.
$$\vec{v}.\frac{\vec{k}}{k} = \frac{k}{2M} + \frac{\omega(k)}{k} \qquad (2.6)$$

It is natural then to define

$$v_c \equiv \min\left(\frac{k}{2M} + \frac{\omega(k)}{k}\right) \qquad (2.7)$$

Clearly if $v < v_c \neq 0$, eq.(2.6) cannot be satisfied since $\vec{v}.\vec{k}/k \leq v < v_c$ and therefore a body moving with velocity $v < v_c$ cannot loose energy by creating an elementary excitation of the fluid. The problem of superfluidity is then reduced to the evaluation of v_c, i.e. to the determination of the energy of the elementary excitations of the fluid as a function of k. For a fluid with elementary excitations $\omega(k) = k^2/2m$, as for the free Bose gas, we get $v_c = 0$ so that for any velocity v eq.(2.6) can be satisfied and we have a normal fluid. On the other hand, fluids with phonon-like excitations with $\omega(k) = ck$ would lead to $v_c = c$ i.e; to a superfluid behaviour. Elementary excitations with a phonon-like spectrum may look rather unlikely for a fluid, since acoustical phonons are typical of rigid systems, like crystals. As we will show in the next section, the liquid helium has such type of excitations, i.e. for low k the system responds "rigidly" to external perturbations. As we will see such rigidity is connected to an "ordering" of the system and the phonon-like excitations correspond to small oscillations in the presence of an order parameter (like the spin waves in ferromagnets).

2.3 Ground state and elementary excitations

Liquid helium consists of helium atoms, which are bosons, interacting with a two-body (short range) potential. The Hamiltonian for the system in a volume V is then of the form

$$H = H_o + g\,H_{int} = \int d^3x\ \phi^*(x)\ (-\frac{\Delta}{2m})\phi(x) + $$
$$+ g \int d^3x\, d^3y\ \phi^*(x)\phi^*(y)\ U(x-y)\phi(y)\phi(x) \qquad (2.8)$$

To discuss the properties of the ground state and of the elementary excitations one has to realize that the separation of the Hamiltonian into $H_o + gH_{int}$ is not very convenient since a perturbative expansion based on H_o leads to divergent terms as $V \to \infty$.(*) These divergences essentially originate because representations labelled by different condensation parameters n_o are inequivalent in the infinite volume limit and H_o leads to an expansion based on a wrong value of n_o. To this effect, the interaction term gH_{int} is not a small perturbation and it is convenient to isolate the terms of H_{int} which are responsible for the above divergences and to include them in a new definition of the unperturbed Hamiltonian. This is essentially equivalent to a partial resummation of the perturbative series and the result is to properly take into account collective (non-perturbative) effects associated to H. To this purpose, one isolates the part of the interaction, (H_{sing}), which is responsible for the $V \to \infty$ divergences to leading orders in g. One then treat $H_o + H_{sing}$ exactly by calculating the correpsonding ground state and

(*) For a detailed discussion of these divergences and for the general strategy see N. Hugenholtz and D. Pines, Phys.Rev. 116 , 489 (1959).

the spectrum of elementary excitations. As we will see, the interaction leads to a complete redefinition of the degrees of freedom with non-perturbative features, like non-analyticity in g. To identify the leading terms of H_{int} it is convenient to work with momentum space canonical variables; then

$$H = \sum\omega(k)\, a^*(k)a(k) + \frac{g}{2V}\sum U(k_2 - k_3)a^*(k_1)a^*(k_2)a(k_3)a(k_4)\,\delta_{k_1+k_2-k_3-k_4} \tag{2.9}$$

As we have seen in the preceding section the operator

$$\lim_{V\to\infty} \frac{1}{V}\int d^3x\, \psi(x) = Q_o$$

commutes with $\mathcal{A}$ and it is a multiple of the identity

$$Q_o = \langle\psi(x)\rangle = \sqrt{n_o}\, e^{i\theta}$$

in any irreducible representation. In the infinite volume limit

$$\frac{a_o}{\sqrt{V}} \to Q_o \tag{2.10}$$

and we can make this substitution (*) in the Hamiltonian (2.9). For simplicity in the following we will put $\theta = 0$; for the general case see the discussion at the end of this section.

The substitution (2.10) has the effect of completely eliminating the $k = 0$ mode from the dynamics or equivalently to neglect the dynamical behaviour of the condensate. As a result

$$N' \equiv \sum_{k\neq 0} a^*(k)a(k)$$

does not commute with H (by terms of order V^{-1}) and therefore the condition of fixed density has to be replaced by the equation

$$\langle N'\rangle / V \equiv n' = n - n_o$$

(*) See N. Bogoliubov, J.Phys. USSR 11, 23 (1947) reprinted in D. Pines The Many-Body Problem, Benjamin 1962.

The above subsidiary condition may be incorporated in the dynamical problem by means of a Lagrange multiplier μ . One then determines the ground state for the modified Hamiltonian

$$\tilde{H} = H_o + g H_{int}(n_o) - \mu(N' - nV + n_o V) \qquad (2.11)$$

or equivalently of the Hamiltonian

$$H' = H_o + g H_{int}(n_o) - \mu N' , \qquad (2.12)$$

where $H_{int}(n_o)$ is the interaction term after the substitution (2.10). The minimization conditions for $<\tilde{H}>/V = \tilde{E}_o/V$, with n fixed, give (*)

$$\mu = \frac{\partial}{\partial n_o}\left(\frac{E'_o}{V}\right) \equiv \frac{\partial}{\partial n_o}\frac{<H'>}{V} \qquad (2.13)$$

so that μ has the meaning of the chemical potential for the system with Hamiltonian H'. Since H and H' differ only by a c-number we may use H' in the following.

We can now discuss the singular terms to leading orders in g. The analysis is based on the weak coupling limit and on the (reasonable) assumption

$$n_k(g) = <N_k>/V \underset{g \to 0}{\to} 0 , \quad k \neq 0 ,$$

(as we have seen in the free Bose gas $n_k(g = 0) = 0$). The above working hypothesis will be used in a self-consistent way, by checking its validity at the end of the calculations. The exact solutions of $H_o + H_{sing}$ will in fact lead to $n_k(g) \sim O(g^{3/2})$ for small g. With this behaviour of $n_k(g)$ we analyse

---

(*) Eq.(2.13) is easily obtained from $\partial(\tilde{E}_o/V)/\partial n_o = 0$, by using the Hellman-Feynman theorem ($\partial E_o/\partial\lambda = <\frac{d}{d\lambda} H>$)

the various terms of H_{int}, (V_{ik} denotes the term in which i creation operators and k destruction operators have survived after the substitution (2.10)):

$$V_{22} = \frac{g}{2V} \sum_{k_i \neq 0} U(q)a^*(k_1)a^*(k_2)a(k_2-q)\, a(k_1+q)$$

is of order $gn_k^2 V$,

$$V_{21} = \frac{g}{2}\left(\frac{n_o}{V}\right)^{1/2} \Sigma\,[U(k_2) + U(-k_1)]\; a^*(k_1)a^*(k_2)\, a(k_1+k_2),$$

$$V_{12} = \frac{g}{2}\left(\frac{n_o}{V}\right)^{1/2} \Sigma\,[U(-k_1) + U(k_2)]\; a^*(k_1+k_2)\, a(k_2)a(k_1),$$

are of order $gn_o^{1/2}\, n_k^{3/2}\, V$,

$$V_{02} + V_{20} = \frac{gn_o}{2}\, \Sigma[U(k)a(-k)a(k) + U(k)a^*(k)a^*(-k)]$$

$$V_{11} = \frac{gn_o}{2}\, \Sigma[\,2U(0) + U(k) + U(-k)]\; a^*(k)a(k)$$

are of order $gn_o n_k V$ and finally $V_{oo} = \frac{g}{2} n_o^2\, U(o)V$. The leading orders in g are therefore given by

$$V_{02} + V_{20} + V_{11} + V_{00} \equiv H_{sing}$$

and the new unperturbed Hamiltonian is

$$\hat{H}_o = H'_o + H_{sing} = \Sigma\left[\,\frac{k^2}{2m} - \mu + gn_o\, U(k) + gn_o\, U(0)\,\right] a^*(k)a(k) +$$

$$+ \frac{gn_o}{2}\, \Sigma\, U(k)\; [a(-k)a(k) + a^*(k)a^*(-k)] + \tfrac{1}{2}\, gn_o^2\, U(0)V \tag{2.14}$$

where we have taken $U(k) = U(-k)$, for simplicity.

To lowest order $\hat{E}_o = \langle \hat{H}_o \rangle = \tfrac{1}{2} gn_o^2 U(0)V$ and therefore by eq.(2.13)

$$\mu = gn_o U(0) \tag{2.15}$$

so that the coefficient of $a^*(k)a(k)$ in eq.(2.14) reduces to

$$\frac{k^2}{2m} + g\, n_o U(k) \equiv \varepsilon(k) \qquad (2.16)$$

It is important to stress that the approximation involved in the derivation of eq.(2.15) is a very delicate one in the sense that going beyond the first order for the determination of μ is not consistent with the neglection of $V_{12} + V_{21}$ in the definition of H_o. $V_{12} + V_{21}$ are in fact of the same order as the terms coming from higher order corrections to eq.(2.15)(*). As a matter of fact, an uncareful treatment of this point would lead to a spurious energy gap above the ground state, whereas there is no energy gap to all orders in g if the perturbation expansion is done properly, as shown by Hugenholtz and Pines.

The determination of the eigenstates of H_o is obtained by performing a Bogoliubov transformation which diagonalizes H_o to the form

$$\hat{H}_o = \sum E(k)\, A^*(k) A(k) + \hat{E}_o \qquad (2.17)$$

The new variables A, A* then define the elementary excitations of the system as a result of the collective effects associated to H_{sing}. To obtain eq.(2.17) we write

$$a(k) = u(k)\, A(k) - v(k)\, A^*(-k)$$
$$a^*(k) = u^*(k) A^*(k) - v^*(k)\, A(-k) \qquad (2.18)$$

and for simplicity we consider even functions of k

$$u(k) = u(|k|) \qquad v(k) = v(|k|)$$

(*) This point has been clarified by N. Hugenholtz and D. Pines, loc.cit.

The canonicity condition for A, A* requires

$$|u(k)|^2 - |v(k)|^2 = 1$$

and the diagonalization condition, i.e. the vanishing of the coefficients of AA and A*A* when the substitution (2.18) is done in eq.(2.14), yields

$$|u(k)|^2 + |v(k)|^2 = \varepsilon(k)/\,[\varepsilon(k)^2 - g^2 n_o^2 U(k)^2]^{1/2},$$

$$2v^*(k)u(k) = gn_o U(k)/\,[\varepsilon(k)^2 - g^2 n_o^2 U(k)^2]^{1/2}. \qquad (2.19)$$

Furthermore, one obtains

$$E(k) = [\varepsilon(k)^2 - g^2 n_o^2\, U(k)^2\,]^{1/2}$$
$$= (\frac{k^4}{4\,m^2} + gn_o\,\frac{k^2\,U(k)}{m})^{1/2} \qquad (2.20)$$

For short range potentials U(0) is finite and non vanishing; moreover U(0) must be positive for stability reasons. Then for low k's

$$E(k) \sim (gn_o U(o)/m)^{1/2}\, k \equiv ck \qquad (2.21)$$

The collective effect has thus lead to a redefinition of the degrees of freedom and a substantial redefinition of the energy spectrum with the appearance of phonon like excitations, for low k's. These excitations are indeed very similar to sound waves and in fact the sound velocity in the fluid, defined by

$$v_s^2 = \frac{\partial p}{\partial \rho} \;,$$

where p is the pressure

$$p = -(\frac{\partial \hat{E}_o}{\partial V})_N = -\frac{\partial}{\partial V}\,(\tfrac{1}{2}\, g\,\frac{N_o^2}{V}\,U(0)) = \tfrac{1}{2}\, gn_o^2\, U(0),$$

coincides with c : $v_s = c$.

It is worthwhile to remark that the delicate cancellation between the term $-g^2 n_o^2 U(k)^2$ in eq.(2.20) and the similar term coming from $\epsilon(k)^2$ is crucial for the low k behaviour, given by eq.(2.21). An incorrect determination of μ (see the discussion after eq.(2.16)) would have spoiled this result.

For large k's, we have

$$E(k) \sim \frac{k^2}{2\,m}$$

since for short range potentials $U(k)/k^2 \to 0$, as k becomes large. One then recovers the ordinary energy spectrum of a non relativistic particle of mass m. In this limit one has

$$|u(k)|^2 \simeq 1 \qquad |v(k)|^2 \simeq (gn_o mU(k))^2/k^4 \ll 1,$$

so that the mixing of a,a* in the definition of A,A* is very small.

For intermediate values of k one experimentally observes a new different type of excitations (rotons) related to the non-linearity of the problem and the simple model discussed above does not account for them(*).

It is clear from eqs.(2.17), (2.18), (2.19) that the representation of the CCR compatible with the dynamics cannot be the Fock representation for a,a*, but the Fock representation for A,A* and the two representations are not unitarily equivalent. The ground state is therefore defined by the equation

$$A(k)|\Psi_o\rangle = 0 \qquad \forall k$$

(*) See A.L. Fetter and J.D. Walecka, Quantum Theory of Many Particle Systems, Mc. Graw-Hill Book, New York 1971, pp. 495-499.

and the excited states are obtained by applying polynomials of A* to

We can now check the validity of the estimate $n_k(g) \sim O(g^{3/2})$ for g small. In fact

$$\Sigma < n_k > \quad = \frac{1}{V} \sum_k < a^*_k a_k > = \frac{1}{V} \sum_k |v(k)|^2$$

$$\xrightarrow[V \to \infty]{} \frac{1}{16\pi^3} \int \frac{d^3k}{E(k)} \left[\frac{k^2}{2m} + gn_o U(k) - E(k) \right]$$

For weak interactions (g << 1) the integral is dominated by small k's and one may therefore replace U(k) by U(0). Then, one gets

$$< n_k > = \frac{1}{4\pi^2} (2mgn_o U(0))^{3/2} \int y^2 dy \left\{ \frac{y^2+1}{(y^4+2y^2)^{1/2}} - 1 \right\} .$$

One can compute also the ground state energy per unit volume

$$\frac{E_\Omega}{V} = \tfrac{1}{2}\, gn_o^2\, U(0) - \frac{g^2 n_o^2 m}{16\,\pi^3} \int d^3k \,\frac{U(k)^2}{k^2} + \frac{8}{15\pi^2} m^{3/2} (gn_o U(o))^{5/2} + o(g^{5/2})$$

which shows the non-analiticity in g.

The case $Q_o = n_o e^{i\theta}$ can be obtained from the one discussed above by means of a gauge transformation

$$u(k) \to e^{i\theta} u(k), \quad v(k) \to e^{i\theta} v(k)$$

which implies

$$A(k) \to e^{i\theta} A(k)$$

Clearly the observable quantities like the ground state energy (but not the ground state wave function) are independent of θ .

III SUPERCONDUCTIVITY

3.1 Properties of superconductors

A superconductor is a metal that, below a critical temperature T_c and for not too high currents, behaves as a perfect conductor. This means that, under these conditions, the conductivity is infinite or that the resistivity is zero; therefore by Ohm's law

$$\vec{E} = \rho\vec{j} = 0 \tag{3.1}$$

the electric field vanishes inside the superconductor. The phenomenon persists if an external magnetic field H is introduced provided that $H < H_c(T)$, $H_c(T)$ being the critical value of H at temperature T. Experimentally

$$H_c(T) = H_c(0) \left[1 - \left(\frac{T}{T_c}\right)^2\right] .$$

The Maxwell equations and eq. (3.1) imply that inside the superconductor

$$\frac{\partial\vec{B}}{\partial t} = - c \operatorname{curl} \vec{E} = 0$$

and therefore the magnetic field inside the superconductor cannot vary with time. Actually, the Meissner effect shows that not only $\partial\vec{B}/\partial t = 0$, but also $\vec{B} = 0$. If we start with a situation in which $\vec{B} = 0$ (no external magnetic field) and $T < T_c$, and then we increase the magnetic field up to $H_c(T)$, since B cannot change inside the superconductor it remains zero there. Thus the B-lines do not enter the superconductor. At $T = T_c$ the metal starts behaving as a normal conductor ($\vec{E} \neq 0$ inside and $\vec{B} \neq 0$) and one gets a non-zero B field inside, if an external magnetic field is applied. Conversely, starting from $T > T_c$, $\vec{B} \neq 0$, if the temperature is lowered, experimentally one finds that at $T = T_c$, B jumps to zero inside the metal (<u>Meissner effect</u>). The transformation just described is thus a reversible transformation and $\vec{B} = 0$ always inside a superconductor ($T < T_c$).

For a large classe of superconductors (below T_c) the specific heat C decreases exponentially

$$C \sim \exp\left[-\Delta_0/K_B T\right] ,$$

a behaviour which indicates an energy gap $\Delta_0 \simeq 2K_B T$. For many superconductors, in fact photon absorption occurs only for energies $\hbar\omega > \Delta_0$.

Another very interesting feature of a superconductor is that the condensation energy ε_c defined as the difference between the ground state energy of the metal in the superconductor state and the ground state energy in the normal (conducting) state is of the order of $10^{-7} - 10^{-8}$ eV per electron ($\sim$ 1 Kelving). Now this energy is much smaller than all the other energy scales of the metal: i) the energy widths in metals are of the order of a few eV, ii) the correlation and/or exchange energies are somewhat smaller but always of this order of magnitude, iii) the electron interactions lead to energy of the order of 1-10 eV and the coupling constant is α ($\sim$ 1/137), iv) the electron phonon interactions have a coupling constant of order one and lead to energies of the order of a few eV. It appears therefore rather puzzling to explain the origin of such a small energy scale and one is actually facing what nowadays would be called a <u>hierarchical problem</u>. Such a hierarchical factor between ε_c and the other energy scales can hardly be explained by an approach based on perturbation theory methods and in fact, as we shall see, it indicates that non-perturbative effects are needed to explain the phenomenon.

3.2 Superconductivity and energy gap

Landau's argument for explaining superconductivity is similar to that discussed for superfluidity. Let $E(\vec{p})$ denote the spectrum of elementary excitations in the superconductor. A flow of electric current in the superconductor is equivalent to an overall velocity $\vec{v}$, i.e. to a shift of momentum $\vec{q}$ common to all the electrons in the superconductor. The ground state energy will then get shifted by $\frac{1}{2}Mq^2$, M = total mass of the electron system. If the source of the current $\vec{j}$ is switched off, the current flow will decrease, provided there are transitions leading to creation of elementary excitations and therefore to an attenuation of the overall momentum $\vec{q}$. The energy momentum conservation for such transitions requires

$$\tfrac{1}{2}Mv^2 = \tfrac{1}{2}Mv'^2 + E(p)$$

$$M\vec{v} = M\vec{v}' + \vec{p}$$

i.e.

$$\vec{v}.\vec{p} = \frac{p^2}{2M} + E(p)$$

The above equation cannot be satisfied if $|\vec{v}|$ is smaller than

$$v_c = \min\left(\frac{p}{2M} + \frac{E(p)}{p}\right) \simeq \min \frac{E(p)}{p} \qquad (3.2)$$

(since M is very large). Therefore, for $v < v_c$ there cannot be current attenuation. For the free electrons gas (see Chap. I)

$$E(p) = \frac{p^2}{2m} - \frac{p_F^2}{2m}$$

and correspondingly one gets $v_c = 0$, i.e. no superconductivity. More generally if there is no energy gap above the Fermi sphere, i.e. $E(p) \to 0$ as $p \to p_F$, then clearly

$$\min \frac{E(p)}{p} = 0 \;, \quad \text{i.e. } v_c = 0 \,.$$

On the other side, an energy spectrum of the form

$$E(p) = \frac{1}{2m} [(p^2 - p_F^2)^2 + 4m^2\Delta^2]^{\frac{1}{2}} \tag{3.3}$$

would lead to

$$v_c = \Delta/p_F$$

As we shall see in the following sections, collective effects associated to electron pair condensation give rise to an energy spectrum of the form(3.3) and therefore to a critical velocity related to the "energy gap" Δ.

3.3 Electron-phonon interaction and electron pairs

A strong indication about the interaction which is responsible for superconductivity comes from the observation that for different superconductors the critical temperature T_c is inversely proportional to the mass M of the lattice ion

$$M^{\frac{1}{2}} T_c \simeq \text{constant}$$

(<u>isotope effect</u>). In fact, a lower T_c corresponds to worse superconducting properties and a higher ion mass implies a stronger lattice rigidity and correspondingly a weaker electron-phonon interaction; thus the isotope effect shows that this interaction is responsible for the modification of the electron spectrum.

To see this more precisely, one may start from the Fröhlich Hamiltonia (see Part A, Sect. 2.4)

$$\begin{aligned} H = \int \varepsilon(p)\psi^{\dagger}(p)\psi(p) + \int \omega(k)a^{\dagger}(k)a(k) + \\ g\int F(k)\psi^{\dagger}(p+k)\psi(p)\,[a(k) + a^{\dagger}(-k)] \equiv H_0 + H_{ep} \quad , \end{aligned} \tag{3.4}$$

($F(k) \sim M^{-1/2}$), and perform a canonical transformation

$$H \to \tilde{H} = e^{-S}\, H\, e^{S} = H_0 + H_{ep} + [H_0, S] + \tfrac{1}{2}\,[[H_0, S], S] + \ldots$$

which eliminates H_{ep} up to second order in g. If S is of order g, it must satisfy

$$H_{ep} + [H_0, S] = 0 \ . \tag{3.5}$$

As we have discussed in Part A, Ch. II, the interaction Hamiltonian H_{ep} is not only responsible for scattering processes but also for persistent effects and in particular it gives rise to an effective electron-electron interaction. The role of the above transformation is to exhibit such effect at the lowest order; it is therefore very similar to the transformation which exhibits the existence of a nuclear potential by the elimination of the pion-nucleon interaction (see Part A, Sect. 2.3; in the present case, however, the transformation does not lead to the exact solution of the model).

A solution of eq. (3.5) is

$$S = g \int [A(k)\ a^{\dagger}(-k) + B(k)a(k)]\ F(k)\ \psi^{\dagger}(p+k)\psi(p)d^3pd^3k = -S^{\dagger}$$

with

$$A = -(\varepsilon(p+k) - \varepsilon(p) + \omega(k))^{-1}, \quad B = -(\varepsilon(p+k) - \varepsilon(p) - \omega(k))^{-1},$$

This leads to the following form of $\tilde{H}$ (up to order g^2)

$$\tilde{H} = H_0 + g^2 \sum_{s,s'} \int d^3p'd^3pd^3k \frac{\omega(k)|F(k)|^2}{[\varepsilon(p'-k)-\varepsilon(p')]^2 - \omega^2(k)} \psi^{\dagger}_s(p+k)\psi^{\dagger}_{s'}(p'-k)\psi_{s'}(p)\psi_s(p') \tag{3.6}$$

(plus terms containing only two fermion operators), where the spin indices s,s' have been spelled out. The second term† describes an effective electron-electron interaction with a potential which is not always positive (in contrast with the Coulomb potential). The interaction becomes negative when $|\varepsilon(p'-k) - \varepsilon(p')| < \omega(k)$ and therefore the above Hamiltonian favours the formation of states with pairs of electrons satisfying the above condition. This can be interpreted as an "attractive" force between electrons being induced by the one-phonon exchange‡.

In conclusion, to the Coulomb interaction one adds the second term of eq. (3.6) to get the effective electron-electron interaction in a superconductor.

---

† The same form of the effective e-e interaction could be derived by computing the one-phonon exchange between electrons (see e.g. Part A, sect. 2.3).

‡ For a more detailed discussion see C. Kittel, Quantum Theory of Solids, J. Wiley 1963, Ch. 8.; G. Rickayzen, Theory of Superconductivity, Interscience Publ. 1965.

In order to get the main features of the electron spectrum one may isolate some "dominant" piece of the above e-e Hamiltonian. The euristic idea† is that i) the two body correlations are the dominant effect for the transition from the normal to the superconductor state and ii) the two body correlations are essentially determined by the electron-electron (elastic) scattering near the Fermi surface. The phase space available for such a process, in which the outcoming electrons have momenta close to the Fermi surface, is a function of the total momentum $\vec{K}$ strongly peaked around $\vec{K} = 0$. Moreover the exchange term of the Coulomb interaction is stronger for parallel spins $(1 + \vec{\sigma}_1 \cdot \vec{\sigma}_2)$ and therefore the "attractive" interaction of eq. (3.6) is less contrasted by the Coulomb interaction if the electron pairs have opposite spins. These considerations motivate the choice of the following "reduced" Hamiltonian, the so-called Bardeen-Cooper-Schrieffer (<u>BCS) Hamiltonian</u>, to discuss superconductivity:

$$H = H_0 + \sum_s \int V(k,p)\psi_s^\dagger(p+k)\psi_{-s}^\dagger(-p-k)\psi_{-s}(-p)\psi_s(p) \tag{3.7}$$

where the subscript s denotes the spin variable. Formally, the expression (3.7) is obtained by replacing the integration with respect to p' in eq. (3.6),(and in the similar expression for the Coulomb interaction), by the condition p' = -p. The potential V(k,p) is therefore given by

$$g^2 \frac{|F(k)|^2\omega(k)}{[\epsilon(-p-k)-\epsilon(-p)]^2-\omega^2(k)} + \frac{4\pi e^2}{k^2+\mu^2} \tag{3.8}$$

---

† See e.g. the discussion by L.N. Cooper, Theory of Superconductivity in <u>The Many-Body Problem</u>, Bergen School of Physics 1961, C. Fronsdal ed., W.A. Benjamin 1962, pp. 42-46.

(see Part A Sect. 2.4).

The approximations involved in the derivation of the BCS Hamiltonian cannot be justified completely and some open problems remain†. It is without doubt, however, that the BCS Hamiltonian represents a deep understanding of the physics of superconductors and constitutes a very reasonable model for superconductivity.‡ In any case, it can be regarded as an effective Hamiltonian description of the significant collective effects which are responsible for superconductivity.

For the following, it is convenient to point out the general properties of the interaction Hamiltonian (3.7). First of all, $V(k,p)$ is real

$$V(k,p)^{*} = V(k,p) \tag{3.9}$$

Secondly $V(k,p)$ is even

$$V(k,p) = V(-k,-p) \tag{3.10}$$

Finally since the Hamiltonian must be hermitian only the symmetric part of $V(k,p)$ contributes to eq. (3.7) and therefore $V(k,p)$ may be taken to be symmetric

$$V(k,p) = V(p,k) \tag{3.11}$$

As we have seen explictly in the case of the electron gas (Part B, Ch. I), the Pauli principle induces a "coupling" between the momentum and the spin of the electron and in fact the above interaction can also be written as

---

† For a discussion see e.g. G. Rickayzen, loc. cit. and J. Bardeen, Nobel Prize speech in Physics Today 26, 41 (1973).

‡ For the agreement between theory and experiment see J. Bardeen and J.R. Schrieffer, Prog. Low Temp. Phys., Vol III, North-Holland Amsterdam 1961.

$$\sum_{s,s'} \int U(k,s;p,s')\ \psi^{\dagger}(k,s)\ \psi^{\dagger}(-k,-s)\psi(-p,-s')\psi(p,s').$$

where

$$\begin{aligned} 2U(k,s;p,s') &= V(k,p)\ \delta_{ss'} - V(-k,p)\ \delta_{-s,s'} \\ &= 2U(-k,s;\ -p,s') \end{aligned} \tag{3.12}$$

In the following the spin variable will not be spelled out in general, so that $\vec{k}$ will stand for $\vec{k}$,s and $-\vec{k}$ for $-\vec{k}$,-s. The old equations (3.9),(3.10) and (3.11) remain valid with the new notation for U defined by eq. (3.12). In addition, eq. (3.12) gives

$$U(k;p) = -U(-k;p) \tag{3.13}$$

3.4 BCS model. Exact solution in the thermodynamical limit

In this section we shall take the BCS Hamiltonian as our starting point to show that the phenomenon of superconductivity is related to electron pair condensation and to display the non-perturbative character of such condensation. In our opinion, the features of this phenomenon transcend the specific example in which they arise and have a more general interest. It may be useful to mention the kind of general problems for which superconductivity may qualify as an illuminating prototype. We already know that a metal is characterized by electron condensation in the $\psi^{\dagger}\psi$ channel ($<\psi^{\dagger}\psi> \sim$ electron density), with a condensation energy of the order of 1 eV, the typical scale of the problem. Hartree-Fock methods (see Ch. I) can be used to obtain the main features of such condensation phenomenon. As we shall see, the attractive piece of the effective Hamiltonian (3.7) gives rise to an additional channel of condensation and the corresponding energy scale is hierarchically suppressed with respect to the previous one; the phenomenon is essentially non-perturbative. The BCS model is therefore instructive for understanding non-perturbative fermion condensation phenomena. Finally we shall get insight on the spectrum of elementary excitations in the presence of such non-perturbative condensation.

To discuss the BCS lodel one may adopt the approximation† by which the problem is reduced to the study of a spin system of the Heisenberg type (quasi spin formalism). This is obtained by projecting the BCS Hamiltonian on the subspace of states (called pair states), for which the electron states $|\vec{k}\,\rangle$, $|-\vec{k}\,\rangle$ are either both occupied or both unoccupied.

As emphasized by Haag, the BCS model can actually be solved exactly in the thermodynamical limit and we shall follow Haag's ideas‡. Putting the system in a box of volume V with periodic boundary conditions, the BCS Hamiltonian takes the following form

$$\begin{aligned} H = \sum_p \left(\frac{p^2}{2m} - \mu \right) c^\dagger(p) c(p) + \\ \frac{g}{4V} \sum_{q,p} U(q,p)\, c^\dagger(q)\, c^\dagger(-q)\, c(-p)\, c(p) \end{aligned} \tag{3.14}$$

where μ has the meaning of a chemical potential (see e.g. Sect. 2.1). The equations of motion are

---

† P.W. Anderson, Phys. Rev. 112, 1900 (1958)
C. Kittel, Quantum Theory of Solids, J. Wiley 1963, Ch. 8
W. Thirring, Commun. Math. Phys. 7, 181 (1968); lectures at the International School of Physics, Mallorca 1968, Plenum Press.

‡ R. Haag, Nuovo Cim. 25, 287 (1962)

$$i \frac{d}{dt} c(p,t) = \frac{p^2}{2m} c(p,t) + \frac{g}{2V} \sum_q U(p,q)\, c(-q,t)\, c(q,t)\, c^\dagger(-p,t)$$

where the properties of the potential, (eq. (3.13)), have been used.

Now the Fourier transform $\Delta(x)$ of the operator

$$\Delta_V(p) \equiv \frac{1}{2V} \sum_q U(p,q)\, c(-q)\, c(q) \qquad (3.15)$$

has the form of a quasi local operator, eq. (2.1) (Ch. II Sect. 2.1), labelled by the variable x

$$\Delta_V(x) = \frac{1}{2V} \int_V U(x,z-z')\, \psi(z)\, \psi(z')\, d^3z\, d^3z' ,$$

In fact, for fixed x, $U(x,z-z')$ goes to zero sufficiently rapidly, as $|z-z'| \to \infty$, in such a way that

$$\lim_{V\to\infty} \frac{1}{V} \int_V U(x,z-z')\, \psi(z')\, d^3z' = 0$$

Therefore

$$\lim_{V\to\infty} [\Delta_V(x), \psi^\dagger(y)] = 0 ,$$

so that in the infinite volume limit $\Delta_V(x)$ commutes with the algebra $\mathcal{A}$ generated by the localized operators $\psi(f)$, $\psi^\dagger(g)$ and it is a c-number in any irreducible representation of $\mathcal{A}$. Thus, in the thermodynamical limit the dynamics gets linearized in each irreducible representation of $\mathcal{A}$, since the equations of motion become

$$i \frac{d}{dt} c(p,t) = \frac{p^2}{2m} c(p,t) + g\, \Delta(p)\, c^\dagger(-p,t)$$

$$\Delta(p) = \text{w-}\lim_{V\to\infty} \Delta_V(p) . \qquad (3.16)$$

In each irreducible representation, the generator of time translation can thus be taken of the form

$$H_{eff} = \sum_p \left(\frac{p^2}{2m} - \mu \right) c^{\dagger}(p)\, c(p) + \frac{g}{2} \sum_p \left[\Delta(p)\, c^{\dagger}(p) c^{\dagger}(-p) + \Delta(p)^*\, c(-p)\, c(p) \right] + C \tag{3.17}$$

with C a constant. The Hamilton H_{eff} can be diagonalized by a Bogoliubov transformation (see Sect. 1.1). For simplicity we consider the case in which $\Delta(p)$ is real and look for a transformation

$$c(p,s) = u(p,s)\, C(p,s) + v(p,s)\, C^*(-p,-s) \tag{3.18}$$

with u, v real and satisfying

$$u(p,s) = u(-p,-s) \ , \quad v(p,s) = -v(-p,-s)$$

The canonicity condition reads

$$u(p,s)^2 + v(p,s)^2 = 1 \ .$$

By combining the above equation with the condition that the coefficient of $C^*(p,s)\, C^*(-p,-s)$ vanishes, one gets

$$u^2(p,s) - v^2(p,s) = \left(\frac{p^2}{2m} - \mu \right) / \left[\left(\frac{p^2}{2m} - \mu \right)^2 + g^2 \Delta(p,s)^2 \right]^{\frac{1}{2}}$$

$$2u(p,s)\, v(p,s) = g\Delta(p,s) / \left[\left(\frac{p^2}{2m} - \mu \right)^2 + g^2 \Delta(p,s)^2 \right]^{\frac{1}{2}} \ ,$$

$$H_{eff} = \sum E(p,s)\, C^*(p,s)\, C(p,s) + E_o \ ,$$

$$E(ps,) = \sqrt{\left(\frac{p^2}{2m} - \mu \right)^2 + g^2\, \Delta(p,s)^2} \ . \tag{3.19}$$

E_o is as yet undetermined since H_{eff} is defined up to a constant. Since

$E(p,s) > 0$, the ground state Ψ_o is clearly defined by the condition

$$C(p,s)\ \Psi_o = 0 \qquad \forall\ p,s \tag{3.20}$$

and the representation is a Fock representation for the new operators C, $C^\dagger$. The spectrum of the elementary excitations is given by eq. (3.18) and it shows an "energy gap" above the Fermi sphere. This provides an explanation of superconductivity, as discussed in Sect. 3.1, provided $\Delta(p,s) \neq 0$.

3.5 Gap equation. Non perturbative effects

The gap function $\Delta(p)$ may be determined by self-consistency. Since it is a c-number it can be calculated on any state and therefore by eqs. (3.18) (3.20)

$$\Delta(p) = \lim_{V\to\infty} \frac{1}{2V} \sum_q U(p,q) < \Psi_o,\ c(-q)\ c(q)\ \Psi_o >$$
$$= \frac{1}{16\pi^3} \int d^3q\ U(p,q)\ u(q)\ v(q)$$

The properties of the potential, eq. (3.13), easily imply

$$\Delta(p,s) = -\Delta(-p,-s) \tag{3.21}$$

By using now the expression for uv in terms of Δ one gets an integral equation for Δ (gap equation)

$$\Delta(p,s) = -\frac{1}{32\pi^3} \sum_{s'} \int d^3q\ U(p,s,q,s') \frac{g\Delta(q,s')}{[(q^2/2m-\mu)^2+g^2\Delta(q,s')^2]^{1/2}} \tag{3.22}$$

It is a non-linear equation having a trivial solution

$$\Delta = 0\ ,$$

which corresponds to the Bogoliubov transformation discussed for the free Fermi gas. This solution describes the normal behaviour of a metal. We are interested in non-trivial solutions; in particular we may look for solutions with a ground state invariant under space inversions. In this case one has $u(p,s) = u(-p,s)$, $v(-p,s) = v(p,s)$ and

$$\Delta(-p,s) = \Delta(p,s) \tag{3.23}$$

i.e. $\Delta(p,s)$ is a function of $|\vec{p}|$. Putting

$$\Delta(p) \equiv \Delta(|\vec{p}|,\ s = +) = -\ \Delta(|\vec{p}|,\ s = -1)\ ,$$

$$\bar{U}(p,q) \equiv (4\pi)^{-1} \int d\Omega_q\ U(\vec{p},\ s = +,\ \vec{q},\ s = +)$$

the gap equation becomes

$$\Delta(p) = \frac{-g}{4\pi^2} \int dq\ q^2\ \bar{U}\ (p,q)\ \frac{\Delta(q)}{[(q^2/2m-\mu)^2+g^2\Delta(q)^2]^{1/2}}$$

In the weak-coupling limit the integral is dominated by $q^2/2m = \mu$, i.e. $q = q_F$, and we get

$$\Delta(p) \simeq -\frac{g}{4\pi^2}\ q_F^2\ \bar{U}(p,q_F)\Delta(q_F) \int_0^K dq\ [(q^2/2m-\mu)^2 + g^2\Delta(q_F)^2]^{-1/2}$$

K is a suitable cutoff which should account for the rough approximation of the original integral in the high q region. For $p = q_F$, we obtain

$$1 = -\frac{g}{4\pi^2}\ q_F^2\ \bar{U}\ (q_F,q_F) \int_0^K dq[(q^2/2m-\mu)^2 + g^2\Delta(q_F)^2]^{-1/2}$$

The above equation has a solution only if

$$g\ \bar{U}\ (q_F,q_F) < 0$$

i.e. if the interaction favours the formation of electron pairs close to the Fermi surface. Under this condition one obtains

$$\log(|g|\Delta(q_F) = -\frac{2\pi^2}{|g|mq_F\bar{U}(q_F,q_F)} + C$$

with C a suitable constant which depends on K. In conclusion

$$|g|\Delta \simeq C_o \exp\left(-\frac{2\pi^2}{|g|mq_F\bar{U}}\right) \qquad (3.24)$$

This formula exhibits the highly non-perturbative character of the gap and explains how a hierarchically suppressed energy scale can arise in a theory in which no ad hoc tuning of the parameters has been made.

A similar essential singularity in g is exhibited by the difference between the energy of the normal ground state and the superconducting ground state.

PART C – SYMMETRY BREAKING PHENOMENA

I SPONTANEOUS SYMMETRY BREAKING

1.1 Infinite degrees of freedom and spontaneous symmetry breaking

For a non-perturbative discussion of spontaneous symmetry breaking, a phenomenon which has recently played a very important rôle in many-body and in elementary particle physics, it is convenient to start with the meaning of symmetries in quantum mechanics.

An exact (or unbroken) symmetry in QM has the property that it maps the states $\underset{\sim}{\Phi}$ (i.e. of the rays of a Hilbert space $\mathcal{H}$) onto themselves

$$T : \underset{\sim}{\Phi} \rightarrow \underset{\sim}{\Phi}' \tag{1.1}$$

in such a way that the transition probabilities are left invariant

$$|< \underset{\sim}{\Phi} , \underset{\sim}{\Psi} >|^2 = |< \underset{\sim}{\Phi}' , \underset{\sim}{\Psi}' >|^2 \tag{1.2}$$

It is an important result due to Wigner† that any transformation law of the rays of a Hilbert space $\mathcal{H}$, which satisfies property (1.2) can be described by an operator U in $\mathcal{H}$, which is either unitary or antiunitary:

$$\Phi' = U \Phi \tag{1.3}$$

---

† E. P. Wigner, Group Theory and its Applications to the Quantum Mechanics of Atomic Spectra, Academic Press, N.Y. 1959

V. Bargman, Journ. Math. Phys. 5, 862 (1964)

This implies that the transformation (1.1) induces a corresponding transformation of the observables or of the canonical variables

$$T : A \to A' = U A U^{-1} \tag{1.4}$$

Clearly such transformation preserves the algebraic relations including the adjoint operation (namely $(AB)' = A'B'$, $A^{\star\prime} = A'^{\star}$ etc.) and in particular the commutation relations. A transformation law of the algebra $\mathcal{A}$ of canonical variables, preserving the algebraic relations is called a $\star$ - automorphism[†] of $\mathcal{A}$. Since the equations of motion in the Heisenberg picture can be read as algebraic relations between elements of $\mathcal{A}$, they are invariant under an automorphism of $\mathcal{A}$. An interesting question is whether any automorphism of $\mathcal{A}$ can be written in the form (1.4) and therefore it defines an exact symmetry of the theory.[††]

For quantum mechanical systems with finite degrees of freedom the answer is always affirmative, i.e. for ordinary QM any automorphism of $\mathcal{A}$, or as it is usually called any symmetry of the equations of motion, defines an exact symmetry. In fact, by definition the canonical variables $\{q,p\}$ and the transformed ones $\{q',p'\}$ obey the same canonical commutation relations and therefore, given an irreducible representation π of $\{q,p\}$, (i.e. a set of metrix elements $<\Psi, P(q,p)\,\Psi>$), the representation π'

---

† For simplicity in the following we will often omit the $\star$.

†† It is perhaps worthwhile to mention that two different representation spaces are involved in the above problem. On one side the algebra itself is a representation space for the given automorphism $\alpha : A \to A'$, which may be represented by unitary matrices on that space. On the other side, there is the Hilbert space of states and the question is whether the given automorphism can be represented by a unitary operator in the space of states.

given by

$$\pi'(P(q,p)) = \pi(P(q',p')) \qquad (1.5)$$

is unitarily equivalent to π by Von Neumann's theorem (Part A, Sect. 1.1). This means that there is a unitary operator U such that

$$< U\Psi, P(q,p)\, U\Psi > = < \Psi, P(q',p')\, \Psi >$$

for any Ψ in the Hilbert space $\mathcal{H}$ of the representation π. Thus the given automorphism is described by a unitary transformation $\Psi \to \Psi' = U\Psi$ in $\mathcal{H}$.

The situation drastically changes for the QM of systems with infinite degrees of freedom since there are inequivalent representations of the canonical commutation relations and therefore a symmetry of the equations of motion may fail to give rise to a transformation law of the states, which preserves the transition probabilities. In this case, one says that the symmetry is spontaneously broken. These words reflect the radical change which has taken place in the last decades in treating approximate symmetries in many-body and in elementary particle theory. In the past, the description of physical system exhibiting approximate symmetries was reduced to the problem of identifying explicit "forces" or "perturbations" responsible for such asymmetric effects. In this perspective, the dynamics was described by a Hamiltonian consisting of a dominant symmetric part plus a small asymmetric term. The progress of the last years has shown that the above strategy is not only unconvenient from a practical point of view, since the existence of asymmetric terms complicates the equations of motion and their identification is somewhat arbitrary, but it is actually unacceptable on general grounds, because it is often impossible[†] to reduce symmetry breaking effects to asymmetric terms in the Hamiltonian. The result is that the dynamics must be defined in terms of a symmetric Hamiltonian and that the symmetry breaking is due to a dynamic instability according to which symmetric equations of motion may nevertheless lead to an asymmetric physical description (see further comments below). As we have seen, such phenomena are possible only for infinite quantum mechanical systems.

1.2 Symmetry breaking condition

From now on we will consider those symmetries, i.e. those automorphisms of A, which leave the Hamiltonian invariant, more precisely, which commute with the time translations[††]. They are the non-relativistic analog of the so called internal symmetries in elementary particle physics, characterized by their commutation with the Poincaré group. For such class of symmetries the occurrence of spontaneous breaking can be discussed more easily.

We have already mentioned that the states of an infinite system are described in terms of (local) elementary excitations of the ground state and that for such description the uniqueness of the ground state is a property which can hardly be dispensed with. Clearly, this does not exclude the existence of more than one inequivalent representation of the algebra A

[†] On of the main difficulties of that program is to maintain its predictivity in the presence of radiative corrections and of renormalization effects.

[††] Such additional property is clearly not required for the invariance of the equations of motion (see e.g. the case of Lorentz transformations).

of the canonical variables, each characterized by a ground state and each corresponding to a different "pure phase" of the infinite system. As discussed in Part A, Sect.1.8, the irreducibility of the representation or the validity of the cluster property require that the ground state is unique in each physically reasonable description of the states of the infinite system. The "degeneracy of the ground state", which is frequently stated as a characteristic feature of spontaneous symmetry breaking should actually be understood as the occurrence of more than one inequivalent representation, each with a unique ground state, formally related to one another by the symmetry, see eq. (1.5).

It is sometimes stated that spontaneous symmetry breaking occurs whenever a symmetric Hamiltonian H has one ground state which is not symmetric. In our opinion such statement requires some caution. For a system with a finite number of degrees of freedom the above possibility can occur and the symmetry of H implies that an asymmetric ground state is degenerate with respect to other ground states which are related to it by the symmetry transformation. However, all the degenerate ground states belong to the same Hilbert space in which the theory is defined and the symmetry is described by a unitary operator, which preserves the transition probabilities. For QM_∞ the non-invariance of the ground state is a condition of spontaneous symmetry breaking provided one considers pure phases and symmetries commuting with time translations. In this case, one can show that if the symmetry is not broken then the ground state is invariant. In fact, if U is the unitary operator which describes the symmetry, the equation

$$H\Psi_0 = 0 \qquad (1.6)$$

implies

$$UHU^{-1}U\Psi_o = HU\Psi_o = 0$$

and therefore by the uniqueness of the ground state

$$U\Psi_o = e^{i\theta}\Psi_o \quad . \tag{1.7}$$

More generally, if α is an automorphism of $\mathcal{A}$

$$\alpha : A \to A' \equiv \alpha(A)$$

which commutes with the time translations, a necessary and sufficient condition for α to describe an exact symmetry, in a representation with unique cyclic ground state, is that all the correlation functions are invariant:

$$< \Psi_o , A'\Psi_o > = < \Psi_o , A\Psi_o > \tag{1.8}$$

for any polynomial A of the canonical variables. In fact, if the symmetry is exact there is a unitary operator U such that

$$A' = UAU^{-1} \tag{1.9}$$

and eq. (1.7) implies (1.8). Conversely, if (1.8) holds the transformation

$$\Psi_o \to \Psi_o \;, \quad \Psi = A\Psi_o \to \Psi' = A'\Psi_o \;, \tag{1.10}$$

is defined on a dense set, by the cyclicity of the ground state and it preserves the scalar products, by eq. (1.8):

$$< \Psi' , \Phi' > = < A'\Psi_o , B'\Psi_o > = < \Psi_o , A'^{\star}B'\Psi_o >$$
$$= < \Psi_o , A^{\star}B\,\Psi_o > = < \Psi , \Phi >$$

The transformation (1.10) is thus described by a unitary operator, which describes the (unbroken) symmetry.

1.3 Spontaneous breaking of continuous symmetries

It is of great interest for many-body and for elementary particle physics to consider those sets of automorphisms of $\mathcal{A}$, which form a Lie group G and which are generated by a conserved current, in the sense that for any polynomial A_t of the (localized) canonical variables at an arbitrary time t the infinitesimal variation

$$\delta A_t = A'_t - A_t$$

can be written as†

$$\delta A_t = i \lim_{R\to\infty} \int_{|x|\leqslant R} d^3x \, [\, j_o(x,t_o), A_t \,] \tag{1.11}$$

where j_o is the density of a conserved current $(j_o, \vec{j})$

$$\partial_o j_o + \mathrm{div}\, \vec{j} = 0 \tag{1.12}$$

Unless otherwise stated, j(x,t) will be assumed to transform covariantly under space-time translations

$$U(\vec{a},a_o)\; j(\vec{x},t)\; U(\vec{a},a_o)^{-1} = j(\vec{x}+\vec{a},\, t+a_o) \tag{1.13}$$

It must be stressed that in general the integral of the commutator

---

† More correctly, the integral should be regularized with a test function $f_R(\vec{x}) = f(|\vec{x}|/R)$, with $f \in C^\infty$ and of compact support and with $f(x) = 1$ in the neighbourhood of $x = 0$. The task of giving a mathematically rigorous meaning to the various expressions entering in our discussion, is beyond the scope of these notes, even if it can be done without serious difficulties. For this and other questions, see the excellent review by J.A. Swieca, Goldstone's theorem and related topics, Cargèse lectures 1969.
It must be stressed that condition (1.11) is required for operators which are essentially localized; eq. (1.11) may fail for operators which do not have some sort of localization.

(1.11) has better convergence properties for $R \to \infty$, than the integral

$$Q_R(t_o) \equiv \int_{|x| \leq R} d^3x \; j_o(x,t_o)$$

In local quantum field theory the convergence of the integral (1.11) is actually guaranteed by locality, as we will see in more detail later.

By using (1.11) the condition of unbroken symmetry, eq. (1.8), becomes

$$\lim_{R\to\infty} < \Psi_o \, , [\, Q_R, A \,] \Psi_o > \; = \; 0 \qquad \forall A \in \mathcal{A} \tag{1.14}$$

This condition is clearly implied by (1.8) and (1.11). Conversely, if eq. (1.14) holds then, on the dense set of vectors of the form $\Psi = A\Psi_o$, $A \in \mathcal{A}$, one can define an operator Q

$$Q\,\Psi = \lim_{R\to\infty} \, [\, Q_R, A \,] \Psi_o$$

which is well defined and hermitean. Furthermore, if $\mathcal{A}$ can be decomposed into irreducible finite dimensional representations of G,Q can be exponentiated and $U(\tau) \equiv \exp iQ\tau$ is the unitary operator which describes the symmetry[†].

1.4 Goldstone's theorem for non-relativistic systems

We consider a continuous symmetry α with the following properties:

1) α commutes with the time translations: $\alpha T_t = T_t \alpha$

2) α is generated by a conserved current:[††]

$$\delta A_t = i \lim_{R\to\infty} \int d^3x \; f_R(x) \, [\, j_o(x,0), A_t \,], \tag{1.15}$$

[†] For details see J.A. Swieca, loc. cit.

[††] Technically the limit has to be understood in the distributional sense with respect to the variable t .

Then, if α is spontaneously broken, i.e. for some A

$$<\delta A>_o = i \lim_R <[Q_R(0), A]>_o \neq 0 , \tag{1.16}$$

the energy spectrum cannot have a gap above the ground state (Goldstone's theorem).

To prove the theorem, we start by noting that as a consequence of 1) and 2):

$$\lim_{R\to\infty} [Q_R(t) , A] = \lim_{R\to\infty} T_t([Q_R(0) , A_{-t}])$$
$$= -i\, T_t\, \delta A_{-t} = -i\, T_t\, T_{-t}\, \delta A = -i\, \delta A$$
$$= \lim_{R\to\infty} [Q_R(0) , A]$$

Thus

$$C_t \equiv \lim_{R\to\infty} <[Q_R(t) , A]>_o = C_o \equiv C \tag{1.17}$$

independent of t .

For the two-point function

$$J_o(\vec{k},\omega) = \int e^{i\vec{k}\vec{x}-i\omega t} <[j_o(x,t) , A]>_o d^3x\, dt \tag{1.18}$$

condition (1.16) gives

$$\lim_{R\to\infty} \int R^3\tilde{f}(R\vec{k})\, e^{-i\omega t} J_o(\vec{k},\omega) d^3k\, d\omega = C$$

and since

$$\lim_{R\to\infty} R^3\tilde{f}(R\vec{k}) = \delta(\vec{k})$$
$$\lim_{\vec{k}\to 0} \int e^{-i\omega t} J_o(\vec{k},\omega) d\omega = C \tag{1.19}$$

A careful analysis of eq. (1.19), in the distributional sense, shows that the limit $\vec{k} \to 0$ can be interchanged with the integral sign and one gets

$$\int e^{-i\omega t}\, J_o(0,\omega)\, d\omega \;=\; C \tag{1.20}$$

i.e.

$$J_o(0,\omega) \;=\; C\,\delta(\omega) \tag{1.21}$$

On the other hand by inserting in eq. (1.18) a complete set of states $|\vec{q}\,,\,\omega_\alpha(\vec{q})>$ labeled by the momentum $\vec{q}$, by the energy $\omega_\alpha(q)$ with respect to the ground state and by the index α ,

$$\begin{aligned} J_o(\vec{k},\omega) &= i\,\mathrm{Im}\, \oint d\alpha \int e^{i(\vec{k}-\vec{q})\vec{x}}\, e^{-i(\omega-\omega_\alpha(\vec{q}))t}\, C_\alpha(\vec{q})\, d^3x\, dt\, d^3q \\ &= i\,\mathrm{Im}\, \oint d\alpha \int e^{-i(\omega-\omega_\alpha(\vec{k}))t}\, C_\alpha(\vec{k})\, dt \end{aligned}$$

where

$$C_\alpha(\vec{k}) \;=\; <\Psi_o|j_o(0,0)|\,\vec{k},\omega_\alpha(\vec{k})><\vec{k},\omega_\alpha(\vec{k})\;|A|\,\Psi_o>.$$

Hence , by eq. (1.21)

$$\lim_{\vec{k}\to 0} J_o(\vec{k},\omega) \;=\; i\,\mathrm{Im}\,\lim_{\vec{k}\to 0} \oint d\alpha\, \delta(\omega-\omega_\alpha(\vec{k}))\; C_\alpha(\vec{k}) \;=\; C\,\delta(\omega) \tag{1.22}$$

It is clear that eq. (1.22) is incompatible with the existence of a gap in the energy spectrum, above the ground state.

To conclude that there are discrete excitations with energy spectrum $\omega_{\bar{\alpha}}(k)$ going to zero as $\vec{k}\to 0$ (the so called <u>Goldstone modes</u>), additional information are needed. For example, that for $\vec{k}\to 0$ only discrete values of the index α contribute to eq. (1.22) and that $C_\alpha(\vec{k})$ is sufficiently regular. In the absence of additional information one can only assert the absence of an energy gap, for $\vec{k}\to 0$.

A condition which ensures the existence of Goldstone modes was specified by J. Swieca[†] in analogy with the relativistic case, but its control appears difficult for non-relativistic systems.

The existence of the isolated point $\omega = 0$ in the energy spectrum of elementary excitations at zero momentum can be proved under the simple condition that the charge density is integrable as commutator[‡]. More precisely the condition is

3) the ground state expectation value of the charge density commutator

$$\langle \Psi_o , [j_o(\vec{x}), A_t] \Psi_o \rangle \qquad , \qquad A \in \mathcal{A}$$

is a finite measure in the $\vec{x}$ variable (possibly after time smearing if necessary).

The above condition, which is obviously satisfied in local field theory, states that the total charge Q_R generates the given group of automorphisms independently of the choice of the smearing function $f_R(x)$.

In any case the result of the theorem is rather strong and there have been attempts to evade this conclusion by appealing to various mechanisms or pathologies which would invalidate the conclusion of the theo-

---

† J. Swieca, Cargése Lectures, 1969. The condition is that $J_o(\vec{k},\omega)$ can be written in the form $G(\vec{k},\omega - E(\vec{k}))$, with G a smooth function in its first variable.

‡ G. Morchio and F. Strocchi, ISAS Report 35/84/EP, April 1984.

rem especially in connection with the superconductivity and the Higgs phenomenon†. In our opinion the problems are deeper than what appears in the literature and will be discussed later. Here, it is worthwhile to mention that condition (1.15) involves the commutator of $j(\vec{x},0)$ with the generic element A of $\mathcal{A}$, in particular with operators at times $t \neq 0$. This cannot be decided on the basis of merely kinematical equations like the CCR's (or the CAR's), since it requires the knowledge of the dynamics. From the equal-time commutator

$$i \lim_{R\to\infty} [\, Q_R(t),\, A_t \,] = \delta A_t \tag{1.23}$$

one can obtain the general (unequal-time) commutator

$$i \lim_{R\to\infty} [\, Q_R(t+\tau),\, A_t \,] = \delta A_t \tag{1.24}$$

provided that

$$\lim_{R\to\infty} [[\, Q_R(t), H \,], A \,] = \lim_{R\to\infty} i\,[\, \frac{d}{dt} Q_R(t), A \,] = 0 \tag{1.25}$$

Eq. (1.25) is in general a stronger property than $\alpha T_t = T_t \alpha$ since it is not guaranteed that the action of α on the (infinite volume) Hamiltonian may be described by $\lim_{R\to\infty} [\, Q_R, H \,]$. To show that eq. (1.23) and eq. (1.25) imply eq. (1.24) it suffices to remark that

$$\lim_{R\to\infty} [Q_R(t+\tau) - Q_R(t),\, A_t \,] =$$

$$= \lim_{R\to\infty} \int_t^{t+\tau} dt' \,[\, \frac{d}{dt'} Q_R(t'),\, A_t \,] = 0$$

---

† We list some of the relevant papers:

T.W. Kibble, Broken Symmetries in Proc. of Int. Conf. Elem. Particles, Oxford 1965, p. 19

J.A. Swieca, loc. cit., pp. 223-224

G.S. Guralnik, C.R. Hagen and T.W. Kibble, Broken Symmetries and the Goldstone's theorem, in Advances in Particle Physics, vol.2, R.L. Cool and R.E. Marshak eds. Interscience 1967, esp. Sect. V, eqs. (5.3), (5.15) (5.16).

R.V. Lange, Phys. Rev. Letters 14, 3 (1965); Phys. Rev. 146, 301 (1965).

By using the continuity equation one gets

$$\lim_{R\to\infty} [\frac{d}{dt} Q_R(t), A] = \lim_{R\to\infty} - \int f_R(x) \operatorname{div} [\vec{j}(\vec{x},t), A] =$$

$$= \lim_{R\to\infty} \int d^3y \ (\vec{\nabla} f)(y) \ [\vec{j}(Ry,t), A] \ R^2$$

and the right-hand side vanishes if

$$\lim_{R\to\infty} |\vec{a}|^2 [\vec{j}(\vec{x}+\vec{a}, t), A] = 0, \qquad \forall\, A \in \mathcal{A}. \qquad (1.26)$$

The above equation holds if the time evolution preserves some sort of locality, i.e. if for any A and $B \in \mathcal{A}$ (A and B may be operators at different times)

$$\lim_{|x|\to\infty} |\vec{x}|^2 [A_{\vec{x}}, B] = 0 \qquad (1.27)$$

with $A_{\vec{x}}$ the $\vec{x}$-translated of A. We will call this property short range asymptotic locality. A sufficient condition for eq. (1.27) is that the interaction is a short range interaction†. For example for an interaction Hamiltonian of the form

$$H = \frac{1}{2m} \int \nabla\psi^\star \nabla\psi \, d^3x + \frac{1}{2} \int d^3x \, d^3y \, \psi^\star(x)\psi^\star(y) \, V(x-y)\psi(y)\psi(x) \qquad (1.28)$$

eq. (1.27) holds if V is a short range potential. As we will see in the next section this difficulties do not arise in local quantum field theory.

† See the discussion in J. A. Swieca, Comm. Math. Phys. 4, 1 (1967)

1.5 Spontaneous breaking of continuous symmetries in local quantum field theory

As discussed in Part A, Ch.III, a local quantum field theory is specified by a representation of the algebra A generated by the polynomials of the local fields $\varphi_\alpha(f)$, $f \in C^\infty$ and of compact support, satisfying the locality condition:

$$[\ \varphi(f)\ ,\varphi(g)\ ] = 0 \quad ,$$

if supp f is spacelike with respect to supp g. We are interested in representations of A with a unique translationally invariant state (vacuum state), which is cyclic with respect to the polynomials of $\varphi_\alpha(f)$.

As before, a necessary and sufficient condition for a symmetry to be exact is that all the correlation functions are invariant. For continuous symmetries generated by a conserved current an additional smearing in time is necessary† due to the singular character of the fields at a point (see Part A, sect. 3.2). The local charges will then be defined by

$$Q_{R,\alpha} = \int d^3x\ dt\ f_R(x)\ \alpha(t)\ j_o(x,t) \tag{1.29}$$

with $f_R(x) = f(|\vec{x}|/R)$ as before and $\alpha(t)$ a C^∞-function of compact support, with

$$\int_{-\infty}^{\infty} \alpha(t)dt = \tilde{\alpha}(0) = 1\ . \tag{1.30}$$

$j_o(x,t)$ is now an element of the local field algebra, i.e. it is a local field. A continuous symmetry is said to be generated by a conserved current if

† At least from a mathematical point of view.

$$\delta A = i \lim_{R\to\infty} [Q_{R,\alpha}, A] \tag{1.31}$$

for any polynomial A of the local fields.

As a consequence of locality, the current conservation now implies that the right hand side of eq. (1.25) is independent of $\alpha(t)$. This property is essentially equivalent to eq. (1.24), i.e. to the t-independence of

$$\lim_{R\to\infty} \langle \Psi_0, [Q_R(t), A] \Psi_0 \rangle .$$

In fact, if α_1, α_2 are two C^∞-functions of compact support satisfying eq. (1.30), then $\alpha \equiv \alpha_1 - \alpha_2$ satisfies

$$\int_{-\infty}^{\infty} \alpha(t)dt = 0$$

and therefore it can be written as

$$\alpha(t) = \frac{d}{dt}\beta(t) \equiv \frac{d}{dt}\int_{-\infty}^{t} \alpha(t')dt'$$

with $\beta(t) \in C^\infty$ and having compact support. Hence

$$\begin{aligned} [Q_{R\alpha_1} - Q_{R\alpha_2}, A] &= [Q_{R\, d\beta/dt}, A] \\ &= \int d^3x\, dt\, \vec{\nabla} f_R\, \beta\, [\vec{j}(\vec{x},t), A] \end{aligned} \tag{1.32}$$

and by locality the r.h.s. vanishes in the limit $R\to\infty$, since, for R sufficiently large, the only points $(\vec{x},t)$ for which $\vec{\nabla} f_R\, \beta$ is different from zero are spacelike, with respect to the (bounded) region in which A is localized.

By using again the property of locality, one can prove that in local quantum field theory the spontaneous breaking of continuous symmetries requires the existence of $\delta(p^2)$ singularities in the two-point function $J_0(p,p_0)$, (see eq. 1.18), i.e. the existence of massless modes (<u>Goldstone's modes</u>). The general proof of the Goldstone's theorem, which covers the case, in which the breaking may occur by means of a composite operator or by the non-invariance of an n-point function, will be given in the next section.

Here, we give the simplified argument, due to Goldstone, Salam and Weinberg[†], which exploits Lorents covariance and spectral condition in the simple case in which the symmetry breaking condition $<\delta A> \neq 0$ is realized by a scalar (local) field $\varphi(x)$. By this we mean that[††]

$$U(a,\Lambda)\,\varphi(x)\,U(a,\Lambda)^{-1} = \varphi(\Lambda x + a)$$

To this purpose we consider the two-point function $< j_\mu(x)\ \varphi(y) >_o$; by inserting a complete set of states with four-momentum p we have

$$< j_\mu(x)\ \varphi(y) >_o = \sum\!\!\!\!\!\!\int d^4p_n\, e^{ip_n(x-y)} < 0|\, j_\mu(0)\,|n><n|\,\varphi(0)\,|\,0 > \tag{1.33}$$

By using the relativistic spectrum condition, $p^2_{(n)} \geqslant 0$, $p^o_{(n)} \geqslant 0$, and the Lorentz covariance we can write

$$< 0|\, j_\mu(0)\,|\, n ><n|\,\varphi(0)\,|\,0 > = p_\mu^{(n)}\ \rho(p^2_{(n)})\ \theta\,(p^2_{(n)})\ \theta\,(p^o_{(n)})$$

where $\theta(x) = 1$ for $x \geqslant 0$, $\theta(x) = 0$ for $x < 0$. If we write

$$\theta(p^2) = \int_0^\infty d\,m^2\,\delta\,(p^2 - m^2)$$

we get[†††]

$$< [\, j_\mu(x)\ ,\ \varphi(y)\,] >_o = i\,\partial_\mu \int_0^\infty dm^2\ \rho\,(m^2) \int d^4p\ \delta\,(p^2 - m^2)$$
$$[\ \theta\,(p^o) - \theta\,(-p^o)\ ]\ e^{ip(x-y)} \tag{1.34}$$

[†] J. Goldstone, A. Salam and S. Weinberg, Phys. Rev. 127 965 (1962)

[††] The validity of this property is reasonable for an elementary field describing a pointlike structure, but it is problematic for compound fields or for fields describing bound states.

[†††] This is the so-called Källen-Lehmann representation of the two point function: G. Källen, Helv. Phys. Acta 25, 417 (1952); H. Lehmann, Nuovo Cimento 11, 342 (1954); A.S. Wightman, Phys. Rev. 101, 860 (1956)

Current conservation yields

$$m^2 \rho(m^2) = 0$$

i.e.

$$\rho(m^2) = a\,\delta(m^2)$$

Then

$$< [\, j_o(\vec{x},0) \,,\, \varphi(0) \,] > \;=\; a\,\delta^3(\vec{x})$$

and the symmetry breaking condition gives

$$a = <\delta\varphi(0)> \;\neq\; 0$$

Thus the Fourier transformation of the two point function (1.33) contains a $\delta(p^2)$ singularity. More precisely, the component of $\varphi(y)\,\Psi_o$ in the subspace generated by $j_\mu(f)\,\Psi_o$ has zero mass $(P^2 = 0)$.

1.6 Proof of Goldstone's theorem in local quantum field theory

In the simplified proof given above, Lorentz covariance seems to be the crucial condition; as a matter of fact it is only indirectly so, because for the two point function Lorentz covariance and relativistic spectrum condition imply[†] locality, which is indeed the basic property for the Goldstone's theorem. We have already stressed this fact in Sect. 1.4 .

By exploiting the property of locality we can indeed obtain a general and rigorous proof of the Goldstone's theorem, covering also the case in which the breaking occurs through the expectation value of a "composite" field.

† R. Jost, The general Theory of Quantized Fields, Ann. Math. Soc. 1965

The argument makes use of the so-called Jost-Lehmann-Dyson (JLD) representation†, which holds for the commutator of two <u>local</u> operators. In particular, if A is a local operator

$$<[\, Q_{R\alpha}, A\,]>_o \;=\; i\int_0^\infty dm^2 \left\{ \int d^3y\, \rho_1(m^2,\vec{y})\, [\int d^4x\, \Delta(\vec{x}-\vec{y},x_o;m^2)\; f_R(\vec{x})\,\alpha(x_o)\,] \right.$$
$$\left. +\int d^3y\, \rho_2(m^2,\vec{y})\, [\int d^4x\, f_R(\vec{x})\,\alpha(x_o)\;\partial_o\,\Delta(\vec{x}-\vec{y},x_o;m^2)\,] \right\} \qquad (1.35)$$

where $\rho_i(m^2,\vec{y})$, $i = 1,2$ are tempered distributions with compact support in the $\vec{y}$ variable as a consequence of locality. $\Delta(\vec{x},x_o;m^2)$ is the Green function of the Klein-Gordon equation

$$\Box f + m^2 f = 0 \quad ,$$

in the sense that the solution $f(\vec{x},t)$ corresponding to the initial data

$$f(\vec{x},\, t{=}0) = \rho_2(\vec{x}) \quad , \quad \frac{\partial f}{\partial t}(\vec{x},\, t{=}0) = \rho_1(\vec{x})$$

can be written in the form

$$f(\vec{x},t) = -\int d^3y\,[\,\rho_1(\vec{y})\;\Delta(\vec{x}-\vec{y},t;m^2) + \rho_2(\vec{y})\;\partial_t\;\Delta(\vec{x}-\vec{y},t;m^2)\,]$$

$$\Delta(x,t;m^2) = \frac{-i}{(2\pi)^3}\int d^4p\;(\theta(p_o) - \theta(-p_o))\;\delta(p^2 - m^2)\; e^{i\vec{p}\vec{x} - i p_o t}$$

---

† R. Jost and H. Lehmann, Nuovo Cimento. <u>5</u>, 1598 (1957)
F. Dyson, Phys. Rev. <u>110</u>, 1460 (1958)
H. Araki, K. Hepp and D. Ruelle, Helv. Acta Phys. <u>35</u>, 164 (1962)
A. S. Wightman, Analytic functions of several complex variables, in Les Houches Lectures 1960.
V. S. Vladimirov, <u>Methods of the Theory of Functions of Several Complex variables</u>, Cambridge, Mass. M.I.T. Press 1966.
The case in which the infrared structure of the theory violates positivity, a situation commonly realized in gauge theories, has been discussed in F. Strocchi, Comm. Math. Phys. <u>56</u>, 57 (1977)

For a rigorous proof of the JLD representation we refer to the papers quoted in the last footnote. Here we report a formal non rigorous derivation†. If $\widetilde{F}(p)$ is a function of the four vector p, vanishing for $p^2 < 0$ we can "decompose" it along the various hyperboloids $p^2 = m^2$

$$\widetilde{F}(p) = \int_0^\infty dm^2\, \widetilde{F}(p)\, \delta\,(p^2 - m^2) \equiv \int_0^\infty dm^2\, \widetilde{\rho}(m^2,p)$$

Clearly the Fourier transform $\rho(m^2,x)$ satisfies

$$(\Box + m^2)\, \rho\,(m^2,x) = 0$$

and therefore

$$\rho(m^2,\vec{x},t) = - \int d^3y\, [\, \rho(m^2,\, \vec{y},\, t{=}0)\, \Delta\, (\vec{x}-\vec{y},t;m^2) +$$

$$+ \rho(m^2,\, \vec{y},\, t{=}0)\, \partial_t \quad (\vec{x}-\vec{y},t;m^2)\,]$$

For $F(x)$ we get a representation of the form (1.35)††.

Now, by the relativistic spectrum condition, the Fourier transformation of the two point function $< j_o(\vec{x},t) A >_o$ vanishing for $p^2 < 0$ (and clearly the same is true for $< A\, j_o(\vec{x},t) >_o$). By the above argument we then get the JLD representation for the commutator $J(\vec{x},t) \equiv < [\, j_o(\vec{x},t), A\,] >_o$.

---

† To make the argument rigorous one has to go to five dimensions

†† For euristic arguments it may be useful to note that formally

$$\rho_2(m^2,\vec{y}) = i \int [\, < 0|j_o(0)|\vec{q},m^2 >< \vec{q},m^2|A|0 > - < 0|A|\vec{q},m^2 >< \vec{q},m^2|j_o|0 >\,]$$

$$\exp\, [\, i\vec{q}\,\vec{y}\,]\ d^3q$$

$$\rho_1(m^2,\vec{y}) = i \int e^{i\vec{q}\cdot\vec{y}}\, d^3q\, q_o [\, < 0|j_o|\vec{q},m^2 >< \vec{q},m^2|A|0 > +$$

$$+ < 0|A|\vec{q},m^2 >< \vec{q},m^2|j_o|0 >\,]$$

Moreover, for $(\vec{x},t)$ spacelike with respect to the (bounded) localization region of A, we have $J(\vec{x},t) = 0$, by locality, so that, for fixed t, $J(\vec{x},t)$ is of compact support in the $\vec{x}$ variable. This implies that also $\rho_i(m^2,\vec{y})$ are of compact support in the variable $\vec{y}$.

The second step for the proof of the Goldstone's theorem is to note that by locality $\rho_i(m^2,\vec{y})$ can be written in the following form†

$$\bar{\rho}_i(m^2,\vec{y}) = \bar{\rho}_i(m^2)\,\delta(\vec{y}) + \vec{\nabla}\cdot\vec{\sigma}_i(m^2,\vec{y}) \tag{1.36}$$

where

$$\bar{\rho}_i(m^2) = \int \rho_i(m^2,\vec{y})\, d^3y$$

and $\vec{\sigma}_i$ are of compact support in the variable $\vec{y}$.

Again by locality the second term in (1.36) does not contribute to the right hand side of eq. (1.35), in the limit of large R, since by integration by parts the differential operator $\vec{\nabla}$ can be shifted to $\Delta(\vec{x}-\vec{y},x_0;m^2)$ and then to $f_R(\vec{x})$. Thus for large R, eq. (1.35) reduces to

$$\int_0^\infty dm^2 \Big\{ \bar{\rho}_1(m^2)\,[\int d^4x\,\Delta(\vec{x},x_0;m^2)\,f_R(\vec{x})\,\alpha(x_0)\,] +$$
$$\bar{\rho}_2(m^2)\,[\int d^4x\,\partial_0\,\Delta(\vec{x},x_0;m^2)\,f_R(\vec{x})\,\alpha(x_0)\,]\,.$$

The integrals in square brackets can be evaluated explicitely and they give

$$\int d^3p\,\tilde{f}_R(\vec{p})\,(2p_0)^{-1}[\,\tilde{\alpha}(p_0) - \tilde{\alpha}(-p_0)\,]_{p_0=\sqrt{\vec{p}^2+m^2}}$$

---

† To prove (1.36) one starts by defining

$$\sigma_i^{(1)}(m^2,\vec{y}) = \int_{-\infty}^{y_1} dy_1' \Big\{ \rho_i(m^2,y_1',y_2,y_3) - \delta(y_1') \int_{-\infty}^{\infty} \rho_i(m^2,y_1'',y_2,y_3)\,dy_1'' \Big\}$$

so that

$$\rho_i(m^2,\vec{y}) = \delta(y_1)\int_{-\infty}^{\infty} \rho_i(m^2,y_1'',y_2,y_3)\,dy_1'' + \frac{\partial}{\partial y_1}\sigma_i^{(1)}(m^2,\vec{y})$$

By iterating this procedure to the variables y_2, y_3 one gets eq. (1.36) and

$$\frac{1}{2}\, i \int d^3p \; f_R(p)\, [\tilde{\alpha}(p_o) + \tilde{\alpha}(-p_o)\,] \Big|_{p_o = \sqrt{\vec{p}^2 + m^2}}$$

respectively. Since the commutator (1.35) does not depend on α in the limit of large R, we can choose $\alpha(x_o) = \alpha(-x_o)$ (real), compatibly with condition (1.30), so that

$$\tilde{\alpha}(p_o) = \tilde{\alpha}(-p_o)$$

and only the second integral survives. For $R\to\infty$, $(f_R(\vec{p}) \to \delta(\vec{p}))$, it becomes

$$- \int_0^{\infty} dm^2 \; \bar{\rho}_2(m^2)\; \tilde{\alpha}(\sqrt{m^2}) \tag{1.37}$$

Now, by the general considerations discussed in the previous section, in the limit of large R the right hand side of eq. (1.35) defines a functional of the test function $\tilde{\alpha}(p_o)$ which depends only on the value of $\tilde{\alpha}$ at the origin, (eq. (1.30)). We have just proved that this functional reduces to the expression (1.37). Thus we have

$$\bar{\rho}_2(m^2) = a\, \delta(m^2)$$

and the symmetry breaking condition guarantees $a \neq 0$†.

† It is worthwhile to note that the proof of the Goldstone's theorem presented here covers also the case of theories with infrared singularities which violate positivity and therefore the proof applies also to local gauge theories (for details see F. Strocchi, Comm. Math. Phys. 56, 57 (1977)).

1.7 Breaking of gauge symmetry in superfluidity

As we have seen in Chap. II of Part B, the system is described by the Hamiltonian (2.8) and the algebra A of canonical variables is generated by polynomials of the fields $\psi(f)$, $\psi(g)^{\star}$. The transformation

$$\alpha : \psi(x) \rightarrow e^{i\gamma} \psi(x) \quad , \quad \gamma \in \mathbb{R} \tag{1.38}$$

preserves the algebraic relations, in particular the CCR and the equations of motion, so that α is an automorphism of A or a symmetry. Furthermore the transformation α leaves the Hamiltonian (2.8) invariant. Since γ is a continuous variable, α defines a continuous symmetry and it is not difficult to see that

$$i \lim_{R\rightarrow\infty} \left[\int d^3x \, f_R(\vec{x}) \, \psi^{\star}(\vec{x},0) \, \psi(\vec{x},0) , \psi(\vec{y},0) \right] \equiv i \lim_{R\rightarrow\infty} [N_R(0) , \psi]$$

$$= -i \;\; \psi(\vec{y},0) \;\; = \;\; \delta \, \psi(\vec{y},0) \tag{1.39}$$

$$i \lim_{R\rightarrow\infty} [N_R(0) , \psi^{\star}(\vec{y},0)] \;\; = \;\; i \, \psi^{\star}(\vec{y},0) \;\; = \;\; \delta \, \psi^{\star}(\vec{y},0) \tag{1.39'}$$

Since the potential is assumed to be of short range, the dynamics preserves the short range asymptotic locality. Therefore the equal-time relations (1.39) (1.39') can be extended to unequal-time equations and the symmetry is generated by the conserved current

$$j_o = \psi^{\star}(x)\psi(x) \qquad , \qquad \vec{j}(x) = i(\vec{\nabla}\psi^{\star}(x)\psi(x) - \psi^{\star}(x)\vec{\nabla}\psi(x))$$

(the continuity equation can be easily checked to follow from the equations of motion induced by the Hamiltonian (2.8)).

The transformation (1.38) is a gauge transformation of the first kind and it is natural to define the gauge invariant subalgebra $A_{inv} \subset A$, with the property that each of its elements is gauge invariant. Since the observables involve an equal number of ψ and $\psi^{\star}$, the observable operators are elements of $\mathcal{A}_{inv}$.

In Chap. II, Part B, the Bose-Einstein condensation, at the basis of the phenomenon of superfluidity, has been characterized by the "order parameter"

$$< \Psi_o , \psi(x) \Psi_o > = \xi_o = \sqrt{n_o} \, e^{i\theta} \tag{1.40}$$

The above equation is a symmetry breaking condition for the symmetry (1.38). Clearly, this is in agreement with what discussed in Chap. II of Part B: the parameters n_o and θ label inequivalent irreducible representations of $\mathcal{A}$ and since the gauge transformation (1.38) would lead to

$$\theta \to \theta + \gamma$$

the symmetry is spontaneously broken in any representation of $\mathcal{A}$ with a unique translationally invariant state.

The assumptions of the Goldstone's theorem are satisfied and one deduces the absence of an energy gap. The phonon-like excitations with energy spectrum $\omega(k) = ck$, for small k, are in fact the Goldstone modes corresponding to the spontaneous breaking of the gauge transformation (1.38).

One may consider the possibility of focusing the attention to the gauge invariant algebra without ever introducing gauge dependent operators like $\psi(x)$, $\psi^{\star}(x)$. From a conceptual point of view, such strategy appears more economical, since all what is needed for the physical description of the system is the knowledge of the expectation values of $\mathcal{A}_{inv}$ over the physical states of the system, i.e. one only needs to know the (physically acceptable) representations of $\mathcal{A}_{inv}$. From a practical point of view, however, the above program looks more difficult than the one followed in

Part B, Chap. II in which such representations where obtained through a formulation based on the gauge dependent fields ψ , $\psi^{\star}$. As a matter of fact the equations of motion are more easily written in terms of the fields ψ , $\psi^{\star}$. The Hamiltonian itself is simply defined in terms of ψ , $\psi^{\star}$ and the dynamics is more easily analysed by using $\psi, \psi^{\star}$ as canonical variables.

Furthermore, as discussed in Part A, sect. 1.8, the "completeness" of the canonical variables essentially guarantees that the ground state is cyclic with respect to the algebra of canonical variables (this is their main "raison d'être"). Therefore a representation π is completely determined by the correlation functions of the gauge dependent variables:

$$< \Psi_o \ , \ \psi(x_1) \ \ldots \ \psi^{\star}(y_1) \ \ldots \ \Psi_o >$$

These properties no longer hold for A_{inv} : physically relevant representations of A_{inv} may in general be reducible with respect to A_{inv} and the ground state may not be cyclic with respect to A_{inv} †.

Finally, by using only gauge invariant operators one looses the possibility of getting general information on the phenomenon of superfluidity by relating it to a symmetry breaking; in particular, the occurrence of phonon-like excitations (Goldstone modes) does not have a simple and general explanation in terms of spontaneously broken symmetry. The gauge transformations become the identity transformation on A_{inv} and clearly there cannot be any symmetry breaking. Inequivalent representations of A_{inv} will still be labeled by the (gauge invariant) parameter n_o but the connection with the symmetry breaking order parameter $< \varphi >$ is lost. We will return to such type of questions when we will discuss gauge theories.

---

† In this case the representation is not simply recovered from the correlation functions of elements of A_{inv} .

1.8 Dynamic instability and spontaneous symmetry breaking

We have discussed general non-perturbative criteria for the occurrence of spontaneously broken symmetries and the natural question is how we know in practice when such a phenomenon occurs. A largely used tool has been essentially the mean field approach. This means that for a theory described by a Lagrangian like

$$\mathcal{L} = -\frac{1}{2}(\partial_\mu \varphi)^2 + V(\varphi)$$

with $V(\varphi)$ a polynomial bounded from below and of order up to four (stability and renormalization constraints), the strategy is to look for the absolute minima of $V(\varphi)$ and pick one of them say $\bar{\varphi}$. Then the theory is (perturbatively) defined by considering the (small) quantum fluctuations around $\varphi = \bar{\varphi}$. This perturbative expansion leads to a vacuum expectation value $\langle \varphi \rangle$ which is equal to $\bar{\varphi}$ at lowest order. Spontaneous symmetry breaking is thus related to the fact that an invariant polynomial $V(\varphi)$ might have a non-invariant absolute minimum $\bar{\varphi}$ and therefore the theory constructed around $\bar{\varphi}$ is not symmetric (Goldstone criterium). Since the mean field approximation sometimes leads to incorrect results a different better criterium is needed. A possibility is provided by the methods† of constructive field theory and by the so-colled functional integral approach. To that purpose, the theory is formulated in euclidean space†† and an ultraviolet

† G. Velo and A. S. Wightman eds., Constructive Quantum Field Theory, School of Math. Phys., Erice 1973, Springer Lecture Notes in Physics, vol. 25, Springer 1973

†† This is obtained by analytically continuing the correlation functions $W(y_1, y_2, \dots y_{n-1})$ to imaginary times (this continuation is made possible by locality and spectrum condition). The so obtained correlation functions are called Schwinger functions and they are covariant under the euclidean group.

(lattice space) cutoff and an infrared (finite box) cutoff are introduced. The problem is then reduced to a statistical mechanical problem, where the occurrence of a phase transition or of a spontaneous symmetry breaking is determined by the dependence of the correlation functions from the boundary conditions, in the infinite volume limit. The basis for this is that† any ("locally Gibbs") state over the algebra A can be obtained by suitably choosing the boundary conditions. The resolution of the theory into pure phases is thus obtained by suitably specifying the boundary conditions. A non-vanishing order parameter can thus be obtained.

An equivalent way of resolving the correlation functions obtained by the functional integral into those corresponding to pure phases is to introduce an external field (typically a linear function of the dynamical variable, e.g. $\epsilon(x)\ \varphi(x)$) and look for the limit $\epsilon \to 0$ after the thermodynamical limit has been taken. The dependence of the correlation functions from the way the limit $\epsilon \to 0$ is taken, indicates a phase transition and/or a symmetry breaking (Bogoliubov criterium). When, in the infinite volume limit, the correlation functions exhibit a dependence on the boundary conditions or on the way the external field is removed, one says that there is a dynamical instability††. As a matter of fact a small external field (small volume effect) or a surface effect (boundary conditions) are enough to induce "transitions" from one phase to another.

For a more detailed discussion of these ideas which are borrowed from Statistical Mechanics we refer the reader to any standard book on Critical Phenomena.

---

† D. Ruelle, Statistical Mechanics, Benjamin 1969

†† See A. S. Wightman, Constructive Field Theory in Fundamental Interactions in Physics and Astrophysics, Coral Grables 1972, Plenum Press 1973.

1.9 Heisenberg ferromagnet. Spontaneous magnetization and spin waves

The Heisenberg Hamiltonian for describing ferromagnetism is

$$H = -J \sum_{i,\delta} S_i^{\alpha} S_{i+\delta}^{\alpha} + h \sum_i S_i^z \qquad (1.41)$$

(see the discussion in Part B, Sect. 1.2), where the index i labels the lattice sites, $i + \delta$ is the nearest neighbor to i, α denotes the x, y, z components of the spin $\vec{S}$ and h is an external magnetic field pointing in the z-direction[†]. It is not difficult to show that, for finite lattice and also in the infinite volume limit, the ground state $\Psi_o^{(h)}$ is characterized by all the spins pointing in the z direction. Such state is invariant under discrete translations and one has a non-vanishing mean magnetization

$$\vec{M} = \langle \vec{S}_i \rangle = \langle \Psi_o^{(h)}, \vec{S}_i \Psi_o^{(h)} \rangle$$

pointing in z-direction.

In the limit of vanishing external field, $h \to 0$, the state with all the spins pointing in the z direction is a lowest energy state and in the irreducible representation of the spin algebra Σ corresponding to this ground state (i.e. such that Ψ_o is a cyclic vector) one has a non vanishing magnetization (spontaneous magnetization). It is also clear that any other state with all the spins pointing in one direction (say $\vec{n}$) is a lowest energy state, when $h = 0$,

(†) If only the terms corresponding to $\alpha = z$ are retained in eq.(1.41) one gets the Ising Hamiltonian, whereas if only the terms with $\alpha = x, y$ are kept one has the X Y model.

and it characterizes a different, actually <u>inequivalent</u> [†], (irreducible) representation of the algebra Σ. Any such state has a different mean magnetization, i.e. a different value of the magnetic <u>order parameter</u>. One can check that finite volume states Ψ_{oV}, Ψ'_{oV} corresponding to all spins being aligned in different directions define disjoint worlds in the infinite volume limit

$$\lim_{V \to \infty} (\Psi_{oV}, \Sigma \Psi'_{oV}) = 0$$

The physical meaning of this equation is that elements of the (norm closed) spin algebra Σ can only induce local perturbations on the infinite volume ground state and they are therefore unable to induce transitions between Ψ_o and Ψ'_o (in the infinite volume limit). One might also consider representations of Σ corresponding to a mixed phase, namely those in which the cyclic state with lowest energy is a mixture of two pure state corresponding to different magnetizations; the representation would then be reducible and the cluster property would fail.

The spin rotations define automorphisms of Σ which are spontaneously broken in each irreducible representation with non zero magnetization. The local generator is the total spin $\vec{S}_V$

$$\vec{S}_V = \sum_{i \in V} \vec{S}_i$$

(†) To see the inequivalence one considers the ergodic means

$$\vec{S}_\infty = \operatorname*{w-lim}_{V \to \infty} \frac{1}{V} \sum_{i \in V} \vec{S}_i$$

which exist because the algebra Σ is asymptotically abelian (actually strictly local, since spin operators at different sites commute). Since $\vec{S}_\infty$ commutes with Σ, it labels inequivalent representations of Σ, and in each representation it is represented by $\vec{M}$.

in the volume V and

$$\lim_{V\to\infty} < [S_V^\alpha , S_i^\beta] >_o = i\epsilon_{\alpha\beta\gamma} < S_i^\gamma >_o \neq 0 \tag{1.42}$$

(spontaneous symmetry breaking).

In the model considered so far the dynamics involves short range (actually only nearest neighbor) interactions so that the Goldstone theorem applies[†]. The Goldstone modes are the so called spin waves, whose energy spectrum $\omega(k)$ goes to zero as $k \to 0$.

The spectrum of such elementary excitations can be computed by introducing the Bose operators a_i, a_i^* through the equations

$$S_i^z = S - a_i^* a_i \qquad S_i^+ = S_i^x + iS_i^y = (2S - a_i^* a_i)^{1/2} a_i$$

$$S_i^- = a_i^* (2S - a_i^* a_i)^{1/2} \tag{1.43}$$

(Holstein-Primakoff transformation) for spin S.

In the approximation in which one considers only excitations corresponding to small perturbations of the ground state we may expand the square root in eqs. (1.43) and retain only first order terms

$$S_i^z \simeq S - a_i^* a_i \qquad S_i^+ \simeq (2S)^{1/2} a_i$$

(†) R.V. Lange, Phys. Rev. 146, 301 (1966); ibid. 156, 630 (1967).

The Hamiltonian then takes the following form

$$H \simeq \sum_{\vec{k}} 2S\,J(1-\gamma_{\vec{k}})\; b^*_{\vec{k}}\, b_{\vec{k}} \equiv \sum_{k} \omega(k) b^*_k\, b_k$$

where $b_{\vec{k}}$ is the Fourier transform of a_i

$$b_{\vec{k}} = \frac{1}{\sqrt{N}} \sum_{i}^{N} e^{i\vec{k}.\vec{x}i}\, a_i$$

and

$$\gamma_{\vec{k}} \equiv z^{-1} \sum_{\delta} e^{i\,\vec{k}.\vec{\delta}}$$

z being the number of nearest neighbors to one site. The operators b_k, b^*_k are called destruction and creation operators for magnons; clearly $\omega(k) \to 0$ as $\vec{k} \to 0$.

II GAUGE THEORIES AND HIGGS PHENOMENON

2.1 General considerations on gauge theories

With the development of approximate symmetries in elementary particle physics, like SU(2), SU(3), SU(6) etc. the problem of explaining the origin of their breaking became an important issue. To this purpose spontaneous symmetry breaking appeared as an elegant and powerful mechanism; however, since no Goldstone boson with the right quantum numbers was known, a lot of effort was devoted to evade the Goldstone's theorem. By the middle sixties the theorem was put on such a firm basis[†] that it was impossible to evade its conclusions unless some of the assumptions were relaxed. A possibility was offered by gauge theories[††]. On the basis of a perturbative expansion based on a semiclassical approximation it was argued that scalar electrodynamics (the so-called abelian Higgs-Kibble model) exhibited a spontaneous breaking, $<\varphi> = \bar{\varphi} \neq 0$, without Goldstone bosons[‡] (Higgs phenomenon). Since this and other peculiar features of gauge theories are largely independent of the specific models, it is worthwhile to ask whether they can be understood in terms of general physical ideas.

---

† D. Kastler, D.W. Robinson and J.A. Swieca, Comm. Math. Phys. 2,108(1966)

†† In some of the early papers the lack of manifest covariance was emphasized as the relevant feature which prevents the proof of Goldstone's theorem. As it has been stressed in the previous sections and it will discussed in detail later, the lack of locality is the crucial issue.

‡ P.W. Higgs, Phys. Letters 12, 132 (1964); Phys. Rev. Letters, 13, 508 (1964); Phys. Rev. 145, 1156 (1966)
F. Englert and R. Brout, Phys. Rev. Letters 13, 321 (1964)
G.S. Guralnik, C.R. Hagen and T.W. Kibble, Phys. Rev. Letters 13, 585 (1966)

The standard characterization of gauge theories is the invariance under a group $\mathcal{G}$ of gauge transformations of the second kind (the so-called local gauge transformations). However this invariance property does not have a direct physical interpretation since gauge transformations act non-trivially only on unphysical quantities. The original motivation by Yang and Mills[†] relates the requirement of gauge invariance to the requirement that internal quantum numbers or charges have only a local meaning and that their relative identification at different points is meaningless. This sort of equivalence principle does not appear to be supported by strong physical arguments and becomes rather puzzling in the cases (confinement and Higgs models) in which the physical states do not carry gauge quantum numbers, (see below). As a matter of fact, in any formulation in which only observable quantities appear, the gauge symmetry collapses to the identity. It is therefore natural to ask whether the crucial features of gauge theories can be traced back to a more physical property than the invariance under unobservable gauge transformations.

To this purpose we start from classical considerations and recall that a gauge group $\mathcal{G}$ is an infinite dimensional Lie group, since the group parameters are functions of space-time. (The subgroup G of $\mathcal{G}$ corresponding to constant group parameters is a standard continuous Lie group). Now, the invariance of the Lagrangian under a discrete group provides only conservation rules; when the group is enlarged to a continuous Lie group G one does not only get a conservation law

$$Q^{\alpha} = \int d^3x \; j_0^{\alpha}(\vec{x},t) = \text{const} \qquad (2.1)$$

for each generator of G but also a continuity equation

† C.N. Yang and R.L. Mills, Phys. Rev. 96, 191 (1954)

$$\partial^{\mu} j^{\alpha}_{\mu}(x) = 0 \tag{2.2}$$

This equation implies that the "charge" associated to j^{α}_{μ} is <u>conserved locally</u> (not only globally), in the sense that for any finite volume V the variation of the charge inside must be completely accounted for by a charge flux through the boundary surface. When the continuous Lie group G is extended to an infinite dimensional Lie group the invariance of the Lagrangian leads to a further strengthening of the conservation laws.

We consider for simplicity the abelian case. In this case the invariance of the Lagrangian $\mathcal{L} = \mathcal{L}(\varphi, \varphi^{\star}, A_{\mu})$ under the gauge transformation

$$\varphi(x) \to e^{ig\,\Lambda(x)}\,\varphi(x) \quad , \quad A_{\mu}(x) \to A_{\mu}(x) + \partial_{\mu}\Lambda(x)$$

yields the following continuity equation

$$\partial^{\mu}\left\{ \frac{\delta\mathcal{L}}{\delta\,\partial_{\mu}\varphi}\, ig\Lambda\varphi + \text{comp. conjug.} + \frac{\delta\mathcal{L}}{\delta\,\partial_{\mu}A_{\nu}} \cdot \partial_{\nu}\Lambda \right\} = 0 \tag{2.3}$$

Since $\Lambda(x)$ is an arbitrary function of space-time the above equation seems to lead to infinite conservation laws. As a matter of fact one gets

$$\partial^{\mu} j_{\mu}\Lambda(x) + j_{\mu}\,\partial_{\mu}\Lambda(x) + \partial_{\mu}\left[\frac{\delta\mathcal{L}}{\delta\partial_{\mu}A_{\nu}}\right]\partial_{\nu}\Lambda(x) + \frac{\delta\mathcal{L}}{\delta\,\partial_{\mu}A_{\nu}}\,\partial_{\mu}\partial_{\nu}\Lambda(x) = 0 \ ,$$

where

$$j_{\mu}(x) \equiv ig\,\frac{\delta\mathcal{L}}{\delta\partial_{\mu}\varphi}\,\varphi + \text{compl. conj.} \ ,$$

is the current which generates constant gauge transformations and the arbitrariness of $\Lambda(x)$ implies

$$\partial^{\mu} j_{\mu}(x) = 0 \tag{2.4}$$

$$j_{\mu} = -\partial^{\nu}\,\frac{\delta\mathcal{L}}{\delta\,\partial_{\mu}A_{\nu}} \equiv -\partial^{\nu}F_{\mu\nu} \tag{2.5}$$

$$F_{\mu\nu} = -F_{\nu\mu} \tag{2.5'}$$

Thus, the apparently infinite conservation laws are actually equivalent to eqs. (2.5), (2.5').

The argument can be repeated for the non-abelian case. As a consequence of the extension of the invariance group G to an infinite dimensional gauge group, for each generator of G one does not only get the conservation law (2.2), but also a local Gauss's law in the sense that the corresponding j^{α}_{μ} is the divergence of an antisymmetric tensor

$$j^{\alpha}_{\mu}(x) = \partial^{\nu} G^{\alpha}_{\nu\mu}(x) \tag{2.6}$$

Eq. (2.6) implies that the current conservation is reduced to a purely geometrical or kinematical fact, since the equations of motion for the field carrying the associated charge are not needed, (the current is in some way superconserved).

The second important consequence of eq. (2.6) is that, due to Gauss's theorem, the charge Q^{α} can be measured by a flux at (spacelike) infinity. Therefore the charge Q^{α} does not depend on the local behaviour of the solutions, but only on their behaviour at spacelike infinity. Since such behaviour is stable under time evolution and local deformations†, in a certain sense eq. (2.6) says that the charge Q^{α} has a topological character.

As we will show in the following sections, most of the peculiar features of gauge quantum field theories can be derived simply from the validity of a local Gauss's law, independently of the specific model and/or of the equations of motion. We want actually to suggest that the local Gauss's law is the basic and primary feature which characterizes elementary particle interactions in a more physical way than gauge invariance, since it does

† See F. Strocchi, Lectures, in Topics in Functional Analysis 1980-81, Scuola Normale Superiore 1982

not require the introduction of unphysical fields, like the vector potential, and their gauge transformations. Gauge invariance can then be regarded as a technical tool to write down Lagrangian functions which authomatically lead to the validity of a local Gauss's law† .

2.2 Local Gauss's law in quantum field theory

The discussion of the previous section, besed on classical arguments can be extended to the quantum case. Along the lines discussed so far, a quantum description of an infinite system requires to fix the algebra of canonical variables or the field algebra and to impose commutation relations as algebraic constraints. To this purpose, we have to say which are the independent dynamical variables to be used to generate our algebra. The procedure of fixing the independent variables, compatibly with the Gauss's law constraint (eq. (2.6)), goes under the name of fixing the gauge.

For simplicity we restrict our discussion to the abelian case. As it is familiar in quantum electrodynamics, in order to set up a well defined evolution problem (Cauchy problem) one introduces the so-called matter fields $\psi(x)$, $\overline{\psi}(x)$ and the vector potential $A_\mu(x)$. The introduction of gauge dependent variables looks unavoidable if the Hamiltonian has to be expressed as a simple local function of the dynamical variables. However, just because one has the kinematical constraint of the Gauss's law ($\partial^i G_{io} = j_o$) the four variables $A_\mu(x)$ are in general not independent.

---

† For a look to gauge theories which emphasizes the local Gauss's law see F. Strocchi, Gauss's law in local quantum field theory, in Field Theory Quantization and Statistical Physics, D. Reidel Publ.1981

We list some of the possible gauge fixing[†].

i) *Coulomb gauge*. It is defined by the condition

$$\mathrm{div}\,\vec{A} = 0 \quad . \tag{2.7}$$

In the free field case eq. (2.7) implies that also $A_o = 0$. In the interacting case A_o is a dependent variable which is however different from zero. The two-point function of A_μ takes the following form

$$< A_i(x)\, A_j(y) >_o \;=\; (\delta_{ij} - \frac{\partial_i \partial_j}{\Delta})\, D(x-y) \quad . \tag{2.8}$$

It is the same form we derived in the quantization of the (free) electromagnetic field, by using creation and annihilation operators for particles (photons) with helicity ± 1. Since the formulation only involves physical degrees of freedom it is called a physical gauge. It is however a <u>non-local gauge</u> in the sense that locality is violated

$$[\, A_i(x) \;,\; A_j(y)\,] \neq 0 \qquad \text{for} \quad (x-y)^2 < 0$$

Therefore the algebra generated by $\psi(f)$, $\overline{\psi}(g)$ and $A_\mu(h)$, with f, g, h localized test functions, is not a local field algebra.

ii) *Evans and Fulton*[††] *gauge*, also called *temporal gauge*. It is essentially defined by the condition

$$A_o = 0 \tag{2.9}$$

The two-point function of A_μ takes the form

$$< A_\mu(x)\, A_\nu(y) >_o \;=\; -\,(g_{\mu\nu} + \partial_\mu \partial_\nu \partial^{-2} - n_\mu \partial_\nu \partial^{-1} - n_\nu \partial_\mu \partial^{-1}) \cdot D(x-y)$$

[†] For a more detailed and rigorous discussion we refer to F. Strocchi and A.S. Wightman, Jour. Math. Phys. <u>15</u>, 2198 (1974)

[††] E. Evans and T. Fulton, Nucl. Phys. <u>21</u>, 492 (1960)

where $\partial \equiv n^{\mu}\partial_{\mu}$ and n^{μ} is a fixed vector of the form $n^{\mu} = (1, 0, 0, 0)$. When n_{μ} is chosen to be spacelike one has the so-called *axial gauge*. The inverse operator ∂^{-1} is defined by multiplication with $(i\, n^{\mu} k_{\mu})^{-1}$ in the Fourier transform. The commutator is not local in time because of the non-local operator ∂^{-1}.

iii) *Gupta-Bleuler or Feymann gauge*[†]. All the four components of A_{μ} are treated as independent and one has

$$< A_{\mu}(x)\ A_{\nu}(y) > \ = \ - g_{\mu\nu} D(x-y) + \partial_{\mu}\partial_{\nu}G(x-y)$$

where D and G are Lorentz invariant functions which satisfy locality. In this gauge locality and covariance are preserved; the price is that unphysical longitudinal modes are allowed to appear in the formulation of the theory. The Gupta-Bleuler gauge can be regarded as the prototype of the local and covariant gauge. It is worthwhile to remark that locality is a regularizing feature of the theory, so that the singularities are less severe[††] than in the non-local gauges and the renormalization procedure[‡] can be carried through in an easier way. This perhaps explains why practically all the calculations of radiative corrections in quantum electro-

---

[†] A detailed presentation of this gauge can be found in any standard text book on quantum electrodynamics. For a discussion of its mathematical aspects and of its general properties see F. Strocchi and A.S. Wightmann, loc. cit.

[††] For the abelian case see K. Symanzik, Lectures on Lagrangian Quantum Field Theory, Desy Report T71/1 and G. Morchio and F. Strocchi, Nucl. Phys. B211, 471 (1983); B232, 547 (1984).

[‡] N.N. Bogoliubov and D.V. Shirkov, Introduction to the Theory of Quantized Fields, Interscience Publ. N.Y. 1959 H. Epstein and V. Glaser, Adiabatic limit in perturbation theory, Lectures at Erice School 1975 on Renormalization Theory, G. Velo and A.S. Wightman eds., D. Reidel 1976.

dynamics have been performed by using local gauges. From a non perturbative point of view the effect of locality is to "decouple" regions which are far spacelike separated and therefore it keeps the time evolution as a local effect. Clearly the infinite volume limit is greatly simplified in a local theory; the definition of the dynamics may even become problematic in the infinite volume limit if the interaction does not have suitable locality properties[†]. Finally it is worthwhile to mention that locality plus relativistic spectrum condition play an important rôle in passing from a quantum field theory in Minkowski space to the so-called euclidean formulation[††], which has proved to be so useful for a non-perturbative approach to quantum field theory.

In the local gauges the Gauss's law constraint, eq. (2.6), does not hold as an operator equation, but only on the states which have a direct physical interpretation. Since eq. (2.6) is a consequence of gauge invariance it should not be a surprise that eq. (2.6) holds only as an expectation value on gauge invariant states.

[†] This may be seen in simple spin models or in many-body systems.

[††] See e.g. J. Glimm and A. Jaffe, Quantum Physics. A functional integral point of view, Springer 1981

2.3 Gauss's law and locality. Charge superselection rule

Since the local Gauss's law implies that the charge associated to j can be measured by a flux at spacelike infinity, it is natural to ask whether a local excitation of the ground state can have an effect on such behaviour at spacelike infinity. In fact, one can prove that if Q_R is the charge associated to a current which obeys a local Gauss's law, then all the local fields have zero charge:

$$\lim_{R\to\infty} [\, Q_{R\alpha}, A \,] = \lim_{R\to\infty} [\, \partial^{\mu} F_{\mu 0}(\vec{x},t)\; f_R(x)\; \alpha(t), A \,]$$

$$= \lim_{R\to\infty} \int d\,x\; \partial_i f_R(x)\; \alpha(t)\; [\, F_{io}(\vec{x},t), A \,] = 0 \qquad (2.10)$$

because only the points $\vec{x},t$ with $|\vec{x}| \sim R$ contribute to the integral and for R large enough they are spacelike with respect to the localization region of A.

Similarly, if Q_R is an unbroken charge associated to a current which obeys a local Gauss' law, one cannot obtain a charged state by applying a local field to the vacuum.

In fact, if $\Psi = A\,\Psi_o$, with A a local field, one has

$$\lim_R \langle \Psi, Q_R \Psi \rangle = \lim_R \langle \Psi, (\partial F)(f_R \alpha) \Psi \rangle$$

$$= \langle A\,\Psi_o, [\, (\partial F)(f_R \alpha), A \,]\, \Psi_o \rangle +$$

$$\langle A^{\dagger} A\,\Psi_o, (\partial F)(f_R \alpha)\, \Psi_o \rangle$$

Now, the first term vanishes by locality as in eq. (2.10). The last term also vanishes essentially because it describes the matrix element of the flux of the electric field at infinity between two chargeless (local) states. This can be proved in general by using the result[†] that if Q_R is an unbroken charge, for any local state Φ, one has

$$\lim_R \langle \Phi, Q_R \Psi_o \rangle = 0 \quad .$$

Clearly, $(\partial F)(f_R \alpha)$ is an unbroken charge, because by eq. (2.10)

$$\lim_R \langle \Psi_o, [(\partial F)(f_R \alpha), A] \Psi_o \rangle = 0$$

(for any local field A) and $A^+ A \psi_o$ is a local state; therefore the above result applies[‡] and one finally gets

$$\lim_R \langle \Psi, Q_R \Psi \rangle = 0$$

It is worthwhile to remark that the above argument only uses the validity of the local Gauss' law in the expectation values of the state Ψ.

Thus, quite generally, to obtain charged states one must use a field algebra which does not satisfy locality. The lack of locality is not only related to the vector potential (see previous section) but it is also unavoidable if one wants to have field variables which generate charged states from the vacuum.

The alternative offered by the local gauges is to weaken the local Gauss' law in the following way

$$j_\mu = \partial^\nu G_{\nu\mu} + A_\mu \qquad (2.11)$$

---

† D. Maison, Nuovo Cimento A11, 389 (1972).

‡ For a more detailed discussion see R. Ferrari, L.E. Picasso and F. Strocchi, Nuovo Cimento 39A, 1 (1977).

with $\mathcal{A}_\mu$ a purely gauge field with vanishing expectation values on the physical states (weak local Gauss's law).

For example in the Gupta-Bleuler gauge

$$\mathcal{A}_\mu = \partial_\mu \partial^\nu A_\nu$$

describes the photon longitudinal modes and it does not affect the physical matrix elements. Indeed, for any two physical states Ψ, Φ the photon transversality condition requires

$$< \Psi , \partial^\nu A_\nu \Phi > = 0 \qquad (2.12)$$

so that

$$< \Psi , j_\mu \Phi > = < \Psi , \partial^\nu G_{\nu\mu} \Phi > . \qquad (2.13)$$

Eq. (2.12) can be turned into a condition which characterizes those states which have a direct physical interpretation and excludes the unphysical longitudinal modes. In the Gupta-Bleuler gauge this is explicitely obtained by characterizing the physical states as those satisfying the so-called Gupta-Bleuler condition

$$(\partial^\nu A_\nu)^+ \Psi = 0 \quad , \qquad (2.14)$$

where $(\partial A)^+$ denotes the positive frequency (or destruction operator) part of the operator ∂A. Since ∂A is a "free" field, (eq. (2.11) implies $\Box \partial A = 0$), the identification of the destruction operator part is well defined.

Thanks to the weak form of Gauss's law, eq.(2.11), in the Gupta-Bleuler gauge one can have local charged fields; physical charged states, i.e. charged states satisfying the Gupta-Bleuler condition (2.14) can be obtained as suitable limits of local charged states [†].

Finally, it is worthwhile to mention that if a symmetry generated by a current which obeys a local Gauss's law is not broken, the corresponding generator does not only define a selection rule but a superselection rule, i.e. it is impossible to prepare a physical state which is a coherent superposition of eigenstates $|q_1>$, $|q_2>$ with different charges, $q_1 \neq q_2$. Coherent superposition means that one can measure the relative phase in the superposition

$$a|q_1> + \ b|q_2> . \tag{2.15}$$

To prove the above statement we first note that a physical state Ψ is completely identified by the results of all possible measurements on it, i.e. by its matrix elements $<\Psi, A\Psi>$, for any observable operator A .

To measure the relative phase in the superposition (2.15) we need an observable A which has non-trivial matrix elements between $|q_1>$ and $|q_2>$, i.e. we must have

$$<q_1|[Q,A]|q_2> = (q_1 - q_2)<q_1|A|q_2> \neq 0 \quad .$$

Now, locality may be a useful but not necessary property for generic field operators (for example for unobservable fields), but it is unavoidable for observable operators, as required by Einstein causality. In fact, by the discussion of Sects. 1.8 and 3.2 of Part A, only localized measurements are possible in our laboratories and therefore only operators which have some localization property may be observable. Causality then forces

---

[†] For a more detailed and rigorous discussion of the general features of the local gauges we refer the reader to J.M. Jauch and F. Rohrlich, The Theory of Photons and Electrons, 2nd expanded edition, 2nd corrected printing, Springer 1980; F. Strocchi and A.S. Wightman, Journ. Math. Phys. 15 2198 (1974); G. Morchio and F. Strocchi, Nucl. Phys. B211, 471 (1983); B232, 574(1984)

the vanishing of the commutator of two localized observables, when the localization regions are relatively spacelike. Thus if A is an observable, A must belong to the algebra generated by local operators[†].

The proof of the electric charge superselection rule becomes then a simple consequence of the above arguments[††]. If A is an observable operator and $|q>$ is a physical state of charge q also $A|q>$ is a physical state and one has

$$< q_1| [Q,A] |q_2 > = \lim_{R\to\infty} < q_1| [Q_R ,A] | q_2 >$$

$$= \lim_{R\to\infty} <q_1| [(\partial F)_{R\alpha},A] |q_2 >.$$

Since $F_{\mu\nu}$ is an observable operator, locality gives

$$\lim_{R\to\infty} [(\partial F)_{R\alpha},A] = 0 \tag{2.16}$$

by the same argument used for eq. (2.10). The electric charge superselection rule was suggested in the early fifties on the basis of physical considerations[‡] and it has been repeatedly questioned[†††]. The above argument

---

[†] This deep physical property has been emphasized and exploited by R. Haag and D. Kastler, Journ. Math. Phys. 5, 848 (1964). For the formulation of quantum field theory in terms of local algebras of observables see also Cargèse Lectures in Physics vol. 4, D. Kastler ed., Gordon and Breach 1970, and the Proceedings of the Int.School "E. Fermi" Varenna 1973, "C^* algebras and Their Applications to Statistical Mechanics and Quantum Field Theory, D. Kastler ed., Soc. Ital. Fis. Bologna 1976.

[††] The rather sketchy argument presented here can be made more rigorous and extended to the non-abelian case; see F. Strocchi and A.S. Wightman, Journ. Math. Phys. 15, 2198 (1974). There one can also find the discussion of delicate points connected with renormalization.

[‡] G.C. Wick, A.S. Wightman and E.P. Wigner, Phys. Rev. 88, 101 (1952)

[†††] See e.g. Y. Aharonov and L. Susskind, Phys. Rev. 155, 1428 (1967) and D. Kershaw and C. Woo, Phys. Rev. Letters, 33, 918 (1974)

shows that it is the consequence of basic properties: the local Gauss's law or Maxwell equations and Einstein causality.

The extension of the above proof to the non-abelian case leads to interesting properties. For example for an unbroken SU(2) gauge theory, the three charges associated to the three generators T_1, T_2, T_3 define superselection rules. Since Q_1, Q_2, Q_3 do not commute, an eigenstate of say Q_3 with non zero charge, is a non trivial coherent superposition of eigenstates of say Q_1; therefore if physical "coloured" states exist they cannot be eigenstates of any colour charge Q_i. They can only be colourless mixtures, i.e. such that the expectation value of any charge Q_i vanishes on such states[†].

2.4 Gauss's law and Higgs phenomenon

The local Gauss's law, which we have emphasized as the characteristic feature of gauge theories, also provides the key for understanding the Higgs phenomenon, where an apparent symmetry breaking is not accompanied by Goldstone particles. As for the superfluid case, the symmetry associated to gauge transformations of the first kind can be given a non-trivial meaning only by introducing gauge dependent operators. Since the choice of the gauge affects the algebra of gauge dependent fields the discussion of a possible symmetry breaking may in general depend on the gauge.

[†] Because of the mixture character of the states, the vanishing of the expectation values of Q_i does not imply the vanishing of the expectation values of $Q_1^2 + Q_2^2 + Q_3^2$.

For a more detailed discussion see F. Strocchi, Phys. Rev. D17, 2010 (1978) Sect. IV.

In the gauges in which the local Gauss's law holds in the strong form the field algebra cannot satisfy locality if we want to reach sectors other than the zero charge sector. As stressed in Chap. I the lack of locality is already enough to prevent the application of Goldstone's theorem and symmetry breaking may occur without Goldstone particles. In fact, perturbation expansions based on mean field approximations exhibit this phenomenon (Higgs phenomenon). However, since perturbation theory and mean field approximation sometimes lead to incorrect results one may wonder whether the phenomenon can be understood on a better basis and perhaps be turned into a theorem in which the crucial ingredients are clearly specified. This can be done by using local (renormalizable) gauges, whereas the situation is more involved in non-local gauges (see following section).

Before proving the theorem we will give a simple non rigorous argument showing that symmetries generated by currents which obey a local Gauss's law cannot be broken according to the Goldstone mechanism. By Goldstone mechanism we mean the saturation of the commutator $< [Q_R, A] >_o$, with $< \delta A >_o \neq 0$, by massless (boson) states, in the limit $R \to \infty$. This saturation implies that such massless boson states $|p>$, $p^2 = 0$, must have non-vanishing matrix elements with $j_\mu(x)|0>$, i.e.

$$< 0 | j_\mu(x) | p > \neq 0$$

If j_μ obeys a local Gauss's law, the above matrix element can be reduced to the divergence of the matrix element

$$< 0 | F_{\mu\nu}(x) | 0 > = e^{ipx} < 0 | F_{\mu\nu}(0) | p > \equiv e^{ipx} f_{\mu\nu}(p) .$$

The transformation properties of a massless spinless state under the Lorentz group and the Lorentz covariance of $F_{\mu\nu}$

$$U(\Lambda) F_{\mu\nu}(x) U(\Lambda)^{-1} = \Lambda_\mu^{-1\rho} \Lambda_\nu^{-1\sigma} F_{\rho\sigma}(\Lambda x)$$

then imply

$$f_{\mu\nu}(p) = \Lambda_{\mu}^{-1\rho} \Lambda_{\nu}^{-1\sigma} f_{\rho\sigma}(\Lambda p) \quad .$$

Thus $f_{\mu\nu}$ transforms as a covariant tensor under the Lorentz group and by general arguments it must have the following form

$$f_{\mu\nu}(p) = (g_{\mu\nu} g(p^2) + p_\mu p_\nu f(p^2)) \theta(p_o) \quad .$$

The antisymmetry of $F_{\mu\nu}$, a property which defines the local Gauss's law, then gives $f = g = 0$ and $f_{\mu\nu} = 0$.

A more convincing argument about the Higgs phenomenon can be given in the local gauges.

Theorem (Higgs phenomenon). In a local quantum field theory, if the symmetry α is generated by a current j_μ which obeys a weak local Gauss's law and it is spontaneously broken, i.e. for some local operator A

$$< \delta A >_o = \lim_{R\to\infty} < [Q_{R\,\alpha}, A] >_o \neq 0 \quad , \qquad (2.17)$$

then: i) the Fourier transform of $< j_\mu(x,t) A >_o$ must contain $\delta(p^2)$ singularities (Goldstone's modes); ii) because of the weak local Gauss's law the Goldstone's modes cannot correspond to physical particles, i.e. these modes cannot show up in the physical spectrum.

Proof. The statement i) has essentially been proved in Sect. 1.6 of the previous section since the proof given there was based on locality and no commitment was made about the mildness of the infrared singularities which may occur in the correlation functions†.

---

† The point was a proof of the JLD representation covering also the cases of infrared singularities which violate positivity (see F. Strocchi, Commun. Math. Phys. 56, 57 (1977)). We refer to that paper also for a more careful handling of the inner products $< \cdot , \cdot >$, which in local gauges do not satisfy positivity.

To prove statement ii) we remark that by eq. (2.16)

$$0 \neq \langle \delta A \rangle_o = \lim_{R\to\infty} \langle [Q_{R\alpha} - (\partial F)_{R\alpha}, A] \rangle \equiv$$

$$\equiv \lim_{R\to\infty} \langle [A_{R\alpha}, A] \rangle$$

where $A_\mu \equiv j_\mu - \partial^\nu F_{\nu\mu}$. If the $\delta(p^2)$ singularities, which account for the symmetry breaking, are due to contributions of massless particle states Ψ_G one would have

$$\langle \Psi_o , A_\mu(x) \Psi_G \rangle \neq 0$$

This shows that Ψ_G cannot be a physical state, since the vacuum is clearly a physical state and by definition of weak Gauss's law the matrix elements of A_μ vanish between physical states.

The above theorem clarifies the conditions for the occurrence of the Higgs mechanism and it characterizes the phenomenon as due to the disappearence of the Goldstone's modes from the physical spectrum as a consequence of the (weak) local Gauss's law. Since no assumption has been made about A, except locality, the theorem covers also the case in which the breaking occurs by a composite field or by a bound state or by a n-point function, with $n > 1$.

Before closing this section we want to mention a choice of gauge (the so called unitary gauge) defined by the following condition†. Let $\bar{\varphi}$ be a point which minimizes the Higgs potential $V(\varphi)$ and let θ^α denote the matrix representation of the generators of the group G, in the representation R_φ furnished by the Higgs field φ. The gauge is fixed by requiring

† This gauge and its properties have been discussed in detail by S. Weinberg, Phys. Rev. D7, 1068 (1973)

the following condition for the field $\varphi(x)$

$$(\theta^{\alpha}\bar{\varphi}, \varphi(x)) = 0$$

i.e. for each point x the component of $\varphi(x)$ in the direction of $\theta^{\alpha}\bar{\varphi}$, vanish for each α.

Clearly in the unitary gauge the applicability of Goldstone's theorem is evaded in a rather drastic way since the gauge fixing explicitely breaks the group G and the Hamiltonian is no longer invariant under G. The symmetry is explicitely broken by hand and there is no genuine spontaneous breaking phenomenon.

2.5 Higgs phenomenon without a symmetry breaking order parameter

The euristic argument, mentioned in the previous section,seems to indicate that in the gauge in which the local Gauss's law holds in the strong form it may be difficult to break a gauge symmetry spontaneously, since the Goldstone mechanism is not available. The ordinary perturbation expansion for the Higgs models is based on a non zero order parameter $\bar{\varphi} = \langle\varphi\rangle$, which is actually put in by hand on the basis of the classical potential, (essentially a mean field approximation), but there is really no guarantee that "symmetry breaking" occurs if one performs a non-perturbative analysis. The problem becomes even more acute if the Higgs mechanism has to occur by some dynamical field or condensate, for which a perturbative treatment is not available.

To have a more dynamical insight on the Higgs mechanism, it is convenient to use the euclidean formulation and the functional integral approach[†] and the ideas of dynamic instability, discussed in the previous Chapter.

By introducing an ultraviolet cutoff, typically formulating the theory on a lattice, the problem is very similar to a statistical mechanical problem. The correlation functions for finite volume V, $<A_1 \ldots A_n>_V$ are obtained by the formula

$$< A_1 \ldots A_n >_V = Z_V^{-1} \int \mathcal{D}\varphi \ldots e^{-A_V} A_1 \ldots A_n ,$$

$$Z_V = \int \mathcal{D}\varphi \ldots e^{-A_V} ,$$

where A_V is the action integral in the volume V. In the infinite volume limit, one gets in general the correlation functions of a mixed phase, i.e. a reducible representation of the field algebra. To get a pure phase one has to properly chose the boundary conditions, i.e. to resolve the possible dynamical instability by specifying suitable boundary values of the fields. The specification of the boundary condition for finite volume V amounts to a redefinition of the action

$$A_V \to A_V + A_S \quad .$$

The occurrence of a non vanishing order parameter with symmetry breaking is actually governed by the boundary term A_S, whose influence on fixed localized regions may be non zero for suitable values of the parameters of the theory.

The above considerations lead to rather simple and strong results if one adopts the Wilson approach to gauge theories , namely if one uses an action invariant under gauge transformations of the second kind. This amounts to dropping the gauge fixing term in the Lagrangian. For example,

---

[†] See e.g. J. Glimm and A. Jaffe, Quantum Physics, A functional integral point of view, Springer 1981.

for the non-abelian Higgs model the euclidean action without gauge fixing reads[†]

$$A = \int \left\{ \frac{1}{4} (F_{\mu\nu})^2 + [(\partial_\mu + ig A_\mu)\varphi]^2 + V(\varphi) \right\} d^4x \qquad (2.18)$$

where φ is a scalar field, $A_\mu(x)$ is a matrix-field corresponding to the representation of the gauge group generators and

$$F_{\mu\nu} = \partial_\mu A_\nu - \partial_\nu A_\mu + g [A_\mu, A_\nu] . \qquad (2.18')$$

The Wilson action A^W is the lattice version of the action (2.18) and it is invariant under local gauge transformations $\alpha_{\lambda(x)}$. One can try to get a non vanishing symmetry breaking parameter $\bar{\varphi} = \langle\varphi\rangle$ by fixing the boundary condition for the functional integral, $\varphi(x) = \bar{\varphi}$ on the boundary of the volume V. For any localized polynomial $P(\varphi)$ one gets

$$\langle \alpha_{\lambda(x)}(P(\varphi))\rangle = \langle P(\alpha_{\lambda(x)}(\varphi))\rangle =$$

$$= Z^{-1} \int \mathcal{D}\varphi \ldots e^{-A_V^W(\varphi,\ldots)-A_S(\varphi)} P(\alpha_{\lambda(x)}(\varphi)) =$$

$$= Z^{-1} \int \mathcal{D}(\alpha_\lambda^{-1}(\varphi)) e^{-A_V^W(\alpha_\lambda^{-1}(\varphi),\ldots)} e^{-A_S^W(\alpha_\lambda^{-1}(\varphi))} P(\varphi) =$$

where a change of variables $\varphi \to \varphi' = \alpha_\lambda^{-1}(\varphi)$ has been made in the last step. Since one can choose a gauge transformation $\alpha_{\lambda(x)}$ such that $\alpha_{\lambda(x)}$ = identity outside the localization region of $P(\varphi)$ and $\alpha_{\lambda(x)} \neq$ identity inside one gets

$$A_S(\alpha_\lambda^{-1}(\varphi)) = A_S(\varphi) \qquad (2.19)$$

and therefore

$$\langle \alpha_{\lambda(x)}(P(\varphi))\rangle = \langle P(\varphi)\rangle \quad . \qquad (2.20)$$

---

† For more details on gauge theories see e.g. L. D. Faddeev and A. A. Slavnov, Gauge Fields. Introduction to Quantum Theory, Reading, The Benjamin Cummings Publ. Co. 1980.

The argument can be easily generalized to any correlation function of A_μ and φ and one concludes that there cannot be any gauge symmetry breaking in Wilson lattice formulation of gauge theories (Elitzur, De Angelis-De Falco-Guerra's theorem†). By eq. (2.20) only the correlation functions which are invariant under gauge transformations of the second kind are non vanishing in the Wilson approach and practically this amounts to considering only a representation of the gauge invariant algebra (or a reducible representation of the gauge dependent field algebra).

The situation has some features in common with the case of superfluidity (see Sect. 1.7 of the previous chapter) and one may wonder whether the absence of gauge symmetry breaking is actually a trivial consequence of the choice of considering only gauge invariant operators, as in the theory of superfluidity. Since in that case the use of gauge dependent canonical variables turned out to be a very useful tool for discussing the dynamical problem, it looks worthwhile to reconsider the problem in the presence of a gauge fixing. The gauge fixing explicitly breaks the gauge transformations of the second kind and, except for the unitary gauge discussed in the previous section, it is in general invariant under the gauge transformations of the first kind. The presence of a gauge fixing then allows non vanishing correlation functions which are globally but not locally gauge invariant. From a practical point of view this is a useful feature, since the conventional continuum approach to gauge theories involves the use of gauge dependent correlation functions in an essential way: they are the building blocks of the perturbative expansion. The calculation of transition amplitudes in terms of Feynman's diagrams and the renormalization procedure are

---

† S. Elitzur, Phys. Rev. D12, 3978 (1975).

G.F. De Angelis, D. De Falco and F. Guerra, Phys. Rev. D17, 1624 1978.

heavily based on Green's functions which are not invariant under local gauge transformations. Clearly for a theory with a gauge fixing the question of the existence of a symmetry breaking order parameter has to be reconsidered anew.

It has been proved[†] that in the temporal gauge one cannot have a symmetry breaking order parameter by an argument which shows that the gauge fixing is not able to yield a sufficiently strong coupling between the boundary and the interior of V in the limit $V \to \infty$ and therefore <u>the correlation functions are independent of the symmetry breaking boundary conditions</u>. The proof can be supplemented by an argument, valid for a large class of gauges, which explains the result as a restoration of symmetry due to disorder effects induced by field configurations with non-trivial topology (<u>point-like defects or instantons</u>).[††]

In conclusion, for this class of gauges the possible dynamical instability, if it exists associated to the Higgs phenomenon, cannot be resolved by a symmetry breaking boundary condition (in contrast with the case of Bose condensation). The absence of symmetry breaking then demands a characterization of the Higgs phenomenon which is different from that usually given in the standard perturbative approach. We recall that the standard

---

[†] J. Frõhlich, G. Morchio and F. Strocchi, Nuclear Physics <u>B190</u> [FS3], 553 (1981)

[††] See the above reference for details. The argument is reduced to the case of low instanton density, a property which can be justified <u>in Higgs's models</u> since due to quantum fluctuations the instanton size is expected to be finite and small, of the order of m_H^{-1}, the typical mass parameter occurring in the Higgs potential, and the instanton density can be estimated to be of the order of $m_H^4 \exp[-c/g^2]$.

approach to the Higgs phenomenon is based on a perturbative expansion around one absolute minimum $\bar{\varphi}$ of the classical Higgs potential (Goldstone's criterium). The symmetry breaking condition is then put in by hand, on the basis of a semiclassical or mean field approximation. Since $V(\varphi)$ is invariant under the gauge group G any other point $\bar{\varphi}'$ of the critical orbit $\{\bar{\varphi}\}$, i.e. any point $\bar{\varphi}' = \alpha\bar{\varphi}$, $\alpha \in G$, is also an absolute minimum of $V(\varphi)$. The selection of one point $\bar{\varphi}$ of the orbit or more generally the occurrence of symmetry breaking requires a non-perturbative justification, which show that a non-vanishing order parameter $\langle\varphi\rangle \simeq \bar{\varphi}$ can be obtained by specifying a suitable boundary condition or by introducing an external field. The non-perturbative results mentioned above show, however, that for a large class of gauges $\langle\varphi\rangle = 0$, independently of the symmetry breaking boundary conditions introduced to define the infinite volume limit of the functional integral and therefore the standard perturbative approach is not justified. This may look rather disturbing especially because in the standard approach the generation of fermion and vector boson masses is governed by the order parameter $\langle\varphi\rangle$. Typically at lowest order one has[†]

$$M_\psi = f\langle\varphi\rangle$$

$$M_W = g^2\langle\varphi\rangle^2$$

and in general the existence of a symmetry breaking order parameter seems to play a rather crucial rôle in the standard approach.

[†] For the experimental successes of the Glashow-Weinberg-Salam model see E.S. Abers and B.W. Lee, Gauge Theories, Physics Reports vol. 9C, N.1. (1973).

2.6 Symmetric picture. Complementarity principle

We will now present a characterization of the Higgs phenomenon, which does not require a symmetry breaking order parameter and therefore it applies also to the gauge in which $<\varphi>=0$.

This approach is based on the concept of critical orbit $\{\bar{\varphi}\} \equiv \{$ set of points of the form $\alpha\bar{\varphi}$, $\alpha \in G$, with $\bar{\varphi}$ an absolute minimum of $V(\varphi)\}$. Whereas the choice of one point $\bar{\varphi}$ is a gauge dependent step, the orbit $\{\bar{\varphi}\}$ is a gauge invariant concept. The residual group $G_{\{\bar{\varphi}\}}$ of the orbit $\{\bar{\varphi}\}$ is defined as the abstract group which is isomorphic to the stability groups $G_{\bar{\varphi}}$, when $\bar{\varphi}$ varies over $\{\bar{\varphi}\}$. The Higgs phenomenon can then be characterized by the existence of critical orbit $\{\bar{\varphi}\}$ with residual group $G_{\{\bar{\varphi}\}}$ smaller than (not equal to) G. Clearly the rôle of the residual invariance group $G_{\bar{\varphi}}$ of the standard formulation is now played by $G_{\{\bar{\varphi}\}}$. Whereas the selection of a point $\bar{\varphi}$ requires a non-trivial coupling between the boundary condition and the interior of the volume V in the infinite volume limit, the selection of a critical orbit $\{\bar{\varphi}\}$ is essentially a volume effect. In the infinite volume limit the functional measure is essentially concentrated on those field configuration which are "close" to the critical orbit (roughly $e^{-A_V(\varphi)} \to 0$ when $V \to \infty$, if φ is not "close" to $\{\bar{\varphi}\}$).

By making reference to the orbit $\{\bar{\varphi}\}$, one may develop a gauge invariant formulation of the Higgs phenomenon. For simplicity we consider the case of total breaking, namely $G_{\{\bar{\varphi}\}}$ = identity. The discussion will easily generalize to the case $G_{\{\bar{\varphi}\}} = U(1)$, which is physically relevant for the Glaskow-Weinberg-Salam $SU(2) \times U(1)$ unified theory. The general case $G_{\{\bar{\varphi}\}} = H$ is more complicated; however since the complications are not related to the Higgs phenomenon, but rather to the problem of describing physical states in the presence of an unbroken "colour" group H , it will not be discussed here[†]. The existence of a gauge-invariant description[††] is based on the following:

Theorem . If $G_{\{\overline{\varphi}\}}$ = identity, for any set of local fields $\{\psi_\alpha(x)\}$ belonging to an irreducible representation R_ψ of G , there is a linear correspondence between the fields of R_ψ and the G-invariant composite local fields $P^{(i)}(\varphi)\cdot\psi = P^{\star}(\varphi)^{(i)}_{\beta}\,\psi_\beta \equiv \Psi^{(i)}(\varphi)$, which are polynomials in the Higgs scalars and linear in the fields ψ . The correspondence is one to one modulo fields of the same form which vanish on $\{\overline{\varphi}\}$.

Lemma. Let $\{\overline{\varphi}\}$ be an orbit of a representation R_H of a compact Lie group G , $\{\overline{\varphi}\}\equiv\{\varphi = R_H(g)\varphi,\ g\in G\}$, then any function $F(\varphi)$ defined on the orbit $\{\overline{\varphi}\}$ and transforming according to a representation R of G (briefly R-covariant):

$$F(R_H(g)\varphi) = R(g)\,F(\varphi) ,$$

is the restriction to that orbit of an R-covariant polynomial.

Proof. Let V_R be the vector space of the functions on the orbit $\{\overline{\varphi}\}$,

---

[†] For the general case see J. Frõhlich, G. Morchio and F. Strocchi, Phys. Letters 97B, 249 (1980); Nucl. Phys. B190 [FS3] 553 (1981).

[††] The use of gauge invariant fields in specific models with scalars in the fundamental representation has been advocated by G.'t Hooft in Quarks and Leptons, Cargèse Lectures 1979, M. Levy et al. eds. Plenum 1980 and by S. Dimopoulos, S. Raby and L. Susskind, Nucl. Phys. B169, 373 (1980). The general case and the proof of the existence of the symmetric picture in general was discussed in the previous reference, where it was shown to be a purely geometrical fact with no conjecture about the dynamics.

which transform according to R. Since G is a Lie group, V_R is a space of continuous functions on $\{\bar{\varphi}\}$. This follows from

$$|F(\varphi_1) - F(\varphi_2)| = |(1 - R(g_{12}))\, F(\varphi_1)| \quad ,$$

where g_{12} is chosen in such a way that $\varphi_2 = g_{12}\,\varphi_1$ and such that $g_{12} \to$ identity, when $\varphi_2 \to \varphi_1$. V_R has dimensions at most equal to the dimensions of the representation R , since an R-covariant function is completely determined by its value on a fixed point $\bar{\varphi} \in \{\bar{\varphi}\}$.

Since the orbit is a compact space, by the Stone-Weierstrass theorem each element of V_R can be uniformly approximated by polynomials $P_n(\varphi)$ restricted to $\{\bar{\varphi}\}$. Actually, only the R-covariant polynomials are sufficient for the approximation. In fact, if $P_n(\varphi) \to F(\varphi)$ on the orbit $\{\bar{\varphi}\}$, the R-covariant polynomials

$$P_n^R(\varphi) \equiv (\mathrm{Vol}(G))^{-1} \int_G dg\, R^{-1}(g)\, P_n(R_H(g)\varphi) \quad , \quad \mathrm{Vol}(G) \equiv \int_G dg \ ,$$

also converge to $F(\varphi)$ on $\{\bar{\varphi}\}$ since

$$\begin{aligned} |P_n^R(\varphi) - F(\varphi)| &= (\mathrm{Vol}(G))^{-1} \left| \int_G dg (R^{-1}(g)\, P_n(R_H(g)\varphi) - F(\varphi)) \right| \leqslant \\ &\leqslant (\mathrm{Vol}(G))^{-1} \left| \int_G dg\, R^{-1}(g)\, [\, P_n(R_H(g)\varphi) - F(R_H(g)\varphi)] \right| \leqslant \\ &\leqslant \sup_{\varphi \in \{\bar{\varphi}\}} |P_n(\varphi) - F(\varphi)| \to 0 \quad . \end{aligned}$$

Since V_R is finite dimensional and obviously contains the vector space $\mathcal{V}_R$ of R-covariant polynomials restricted to $\{\bar{\varphi}\}$, (the elements of $\mathcal{V}_R$ are actually the equivalence classes with respect to the property of being equal on $\{\bar{\varphi}\}$) , the two vector spaces must coincide.

Proof of the Theorem. $(G_{\{\bar{\varphi}\}}$ = identity)

Any G-invariant local field, constructed in terms of $\varphi \in R_H$ and linear in $\psi \in R$ is of the form of a scalar product $(F(\varphi),\psi)$ with $F(\varphi)$ transforming according to R. Therefore, by fixing a point $\overline{\varphi}$ of the orbit, the above invariant yields a definite component of R. Conversely, let us fix a (normalized) vector $v_i \in R$. For any point $\varphi \in \{\overline{\varphi}\}$, since $G_{\{\overline{\varphi}\}}$ is the identity, there is exactly one element $g_\varphi \in G$ such that

$$R_H(g_\varphi)\,\overline{\varphi} = \varphi \ .$$

Clearly if $\varphi' = R_H(h)\,\varphi$, then $g_{\varphi'} = h\, g_\varphi$. We then define

$$F(\varphi) \equiv R(g_\varphi)v_i \qquad .$$

Thus

$$F(R_H(h)\varphi) = R(g_{R_H(h)\varphi})v_i = R(h\, g_\varphi)v_i =$$

$$= R(h)\; R(g_\varphi)v_i = R(h)\; F(\varphi) \ ,$$

i.e. $F(\varphi)$ transforms according to the representation R. By the above Lemma $F(\varphi)$ is the restriction of a polynomial $P(\varphi)$ to the orbit $\{\overline{\varphi}\}$ and $(P(\varphi),\psi)$ is a G-invariant local field which reduces to $\psi_i \equiv (v_i,\psi)$ when $\varphi = \overline{\varphi}$.

For a simple illustration see the SU(2) model discussed below.

The geometric meaning of the above theorem is that the number of G-invariant composite fields which are polynomials in φ and linear in ψ and are linearly independent on the orbit $\{\overline{\varphi}\}$is exactly equal to the dimensions of R_ψ. One may thus describe the physical particles of R_ψ by gauge invariant fields (symmetric picture). The absence of degenerate multiplets in a situation in which all the correlation function are G-invariant can thus be explained by the fact that physical states are described by composite fields which are not related to one another by transformations of G . To illustrate these features one may show that the fermion mass matrix can be written in a gauge invariant form, without involving a gauge dependent symmetry breaking parameter $<\varphi>$. To this purpose one considers the fermion-Higgs Yukawa coupling $f \, \overline{\psi}_\alpha \psi_\beta \Gamma^{\ell}_{\alpha\beta} \varphi_\ell$, Γ being coupling matrices. In the standard picture the fermion mass is identified with the expectation value of $\Gamma^{\ell}_{\alpha\beta} \varphi_\ell$, in lowest order. This can be seen by expanding the functional integral computation of the two-point function $<\overline{\psi}_\alpha(x) \psi_\beta(y)>$, around $\varphi = \overline{\varphi} = <\varphi>$. In the symmetric picture one must compute the two point function of the fermion gauge invariant fields $<\overline{\Psi^i}(x)\Psi^j(y)>$. To this purpose it is convenient to rewrite the Yukawa term in a different form; we note that the polynomials $P^{(i)}(\varphi)$ can be suitably normalized in such a way that

$$P^i_{\alpha\beta}(\varphi) = P^i_\alpha(\varphi)\, P^i_\beta(\varphi)^\star$$

are G-covariant orthogonal projections satisfying

$$\sum_i P^i_{\alpha\beta}(\varphi) = \delta_{\alpha\beta}$$

Then

$$\overline{\psi}_\alpha \psi_\beta \Gamma^{\ell}_{\alpha\beta} \varphi_\ell = \sum_{ij} \overline{P^i_{\alpha\gamma} \psi_\gamma} \, P^j_{\beta\delta} \psi_\delta \Gamma^{\ell}_{\alpha\beta} \varphi_\ell$$

$$= \sum_{ij} \overline{\Psi^i} \, \Psi^j P^i_\alpha(\varphi)^\star P^j_\beta(\varphi) \Gamma^{\ell}_{\alpha\beta} \varphi_\ell$$

The operator

$$M^{ij} = P^i_\alpha(\varphi)^\star P^j_\beta(\varphi) \Gamma^{\ell}_{\alpha\beta} \varphi_\ell$$

is clearly gauge invariant and its expectation value (which may be non zero even if there is no symmetry breaking order parameter) can be identified with the fermion mass matrix.

The above analysis can be extended to the case in which $G_{\{\overline{\varphi}\}} = U(1)$, which is realized in the $SU(2) \times U(1)$ Glashow-Weinberg-Salam model. In this case the physical particles which in the standard picture are neutral with respect to the residual U(1) group are now described by G-invariant composite fields, as before, whereas the physical particles which have a non zero residual U(1) charge are described by composite fields, which are G-invariant modulo U(1) transformations. More precisely to each charged particle one associate a representation $\{\Psi_\alpha\}$ of G, which is equivalent to R and it is of the form

$$\Psi_\alpha = P_{\alpha\beta}(\varphi)\, \psi_\beta$$

with $P_{\alpha\beta}(\varphi)$ a polynomial of the Higgs field $\bar{\varphi}^\dagger$. The important point is that different charged fields of R_ψ are now replaced by different representation $\{\Psi_\alpha\}^i$ of G. In this way one is reduced to a gauge invariant picture, similar to the case in which local gauge transformations are unbroken†. The correlation function are computed by constructing the bilocal fields or strings

$$\overline{\Psi_\alpha(\varphi)(x)}\ P(\exp \int_{\gamma_{xy}} A_\mu(\xi)\ d\xi^\mu)_{\alpha\beta}\ \Psi_\beta(\varphi)(y)$$

From a practical point of view a simpler description is obtained by associating to each charged field of the standard picture a composite field which is G-invariant "modulo U(1) transformations". The neutrino is described by the G-invariant field

$$\Psi^\nu = iN\varphi\sigma_2\psi = \det(\varphi\psi)$$

with N a normalization constant, $N^2 = \langle\varphi^\star\varphi\rangle^{-1}$. The left-handed electron is described by $\Psi^{e_L} = N\varphi^\star\cdot\psi$ and the right-handed election by $\Psi^{e_R} = \psi_R$.

† For more details see J. Fröhlich, G. Morchio and F. Strocchi, loc.cit.

Thus one has

$$P^1_\alpha(\varphi) = N\varphi_\alpha \quad , \quad P^2_\alpha(\varphi) = N\epsilon_{\alpha\beta}\,\varphi^\star_\beta$$

and

$$< M^e(\varphi) > = Nf \; <\varphi^\star_\alpha \delta^\ell_\alpha \varphi_\ell> = f \; <\varphi^\star \varphi>^{1/2}$$

$$< M^\nu(\varphi) > = Nf \; <\epsilon_{\alpha\beta}\,\varphi_\beta \delta^\ell_\alpha \varphi_\ell> = 0$$

in agreement with the predictions of the standard model (the philosophy in however very different!). Similar computations can be done for the W mass. These results show that the existence of a symmetry breaking order parameter is not necessary and that the experimental successes of the Glashow-Weinberg-Salam model can be explained also if $<\varphi> = 0$.

Finally it is worthwhile to mention that the above proof of the existence of a complete set of gauge invariant fields does not require any conjecture about the dynamics. The existence of a symmetric picture for the Higgs phenomenon also sheds light on the rigorous results of lattice gauge theory [†], which show that in an SU(2) gauge theory with scalars in the fundamental representation no sharp boundary separates the confined and the Higgs regime. This result has been extended to a general group[††] and this feature has lead to the conjecture[†††] of the so-called complementarity

[†] K. Osterwalder and E. Seiler, Ann. Phys. 110, 440 (1978); see also E. Fradkin and S. Shenker, Phys. Rev. D19, 3682 (1979)

[††] For a general gauge group the argument requires the introduction of as many Higgs fields as necessary to completely break the symmetry. This very artificial duplication of Higgs fields has some unpleasant features, like that of generating pseudo Goldstone bosons, and it does not seem reasonable for realistic models.

[†††] S. Dimopoulos, S. Raby and L. Susskind, Nucl. Phys. 173B, 208 (1980).

principle. This principle states that a confining theory of fermions and gauge bosons can be analyzed as if a dynamical Higgs phenomenon takes place (Higgs's picture), or as if the gauge symmetry is unbroken (symmetric picture), the observable results being the same. The group theoretical proof discussed above may be regarded as a proof of the complementarity principle: on one side a gauge invariant description is proved to exist in both the Higgs and the confinement regime, independently of the dynamical behaviour of the theory, and on the other side the phase structure of the theory is reduced to the properties of the residual group. Moreover for a complete description in terms of gauge invariant (composite) fields the above theorem shows that the relevant point is the triviality of the residual group, rather than the condition that the scalars are in the fundamental representation of the group.

EXERCICES

Problem 1 Discuss the difficulty of formulating a relativistic dynamics of (point) particles based on the concept of force at a distance.

(For a discussion see e.g. V.V. Molotkov and I.T. Todorov, Commun. Math. Phys. 79, 111 (1981).

Problem 2 Show that if in a given representation of $\mathcal{A}$, the number operator $N = \sum a_i^* a_i$ is a well (densely) defined operator, then N has a purely discrete spectrum with eigenvalues $\{0,1,2, \ldots\}$ (and in particular there is a "no particle" state).

Problem 3 Show that the shift transformation for the creation and destruction operators

$$a_i \rightarrow a_i + f_i \equiv A_i$$

leads from a Fock representation for a_i to a non-Fock representation for A_i unless $\sum |f_i|^2 < \infty$

(Hint: Compute the expectation value of the two number operators $N_a = \sum a_i^* a_i$ and $N_A = \sum A_i^* A_i$).

Problem 4   Show that the free Hamiltonian $H_o = \sum \omega_k a_k^* a_k$ with $\omega_k \sim |\vec{k}|$ for small $\vec{k}$'s (no mass gap) can be well defined also in representations which are not unitarily equivalent to the Fock representation (Hint: Take $a_k = b_k - f_k$ with b_k in a Fock representation and $f_k \sim k^{-3/2}$ for small k (infrared representations)).

Problem 5   Compute the mean number of phonons in a polaron state in the extreme non-relativistic limit ($|\vec{q}| << 1$) to the first non-trivial order in perturbation theory.

Problem 6   Show that in a quantum field theory with a unique translationally invariant state Ψ_o the two point function

$<\Psi_o, \phi(x)\,\phi(y)\,\Psi_o> = b$, implies that $\phi(x)\,\Psi_o = \sqrt{b}\,\Psi_o$

Problem 7   Compute the mean square deviation from total condensation, $<(N - N_o)^2>_o$ (depletion), for a superfluid Bose system in the Bogoliubov approximation.

Problem 8   Compute the (spatial) size of a Cooper pair in a superconductor using the BCS model.

<u>Problem 9</u> Discuss the spontaneous breaking of the Galilei group in a. condensed system (with non-zero average density)

(Hint: Compute the infinitesimal variation of the current operator $\vec{j}$)

<u>Problem 10</u> Discuss the spontaneous breaking of the electron charge without Goldstone bosons and with energy gap , in the BCS model of superconductivity. Identify a relevant general mechanism which explains how the conclusions of the Goldstone theorem are evaded.

(Hint: Consider e.g. $\lim_{R \to \infty} \left[\frac{d}{dt} Q_R(t), \; \psi\psi \right]$)

<u>Problem 11</u> Discuss the low $\vec{k}$ limit of the energy spectrum of a charged Bose gas using eq.(2.20) of Part. B ($U(k) = 4\pi/k^2$).

<u>Problem 12</u> Show that for an isotropic ferromagnet with nearest neighbours interactions

$$H = -J \sum_{i,\delta} S_i^{\alpha} \, S_{i+\delta}^{\alpha}$$

and periodic boundary conditions,ground states have all the spins aligned. (Hint: Use the identity $\vec{S}_i \cdot \vec{S}_i = \frac{1}{2}(\vec{S}_i + \vec{S}_j)^2$ + c-number, and that by angular momentum addition rules the highest eigenvalue for $(\vec{S}_i + \vec{S}_j)^2$ is obtained if the spin $\vec{S}_i$ and $\vec{S}_j$ are aligned).

Problem 13   Apply the Goldstone theorem to spontaneous magnetization in a ferromagnet.

(See e.g. R.V. Lange Phys. Rev. 146, 301 (1966); ibid. 156, 630 (1967))

Problem 14   Find the gauge invariant form of the vector bosons mass operator in the Glashov-Weinberg-Salam model.

(Hint: Use the symmetric picture).

www.ingramcontent.com/pod-product-compliance
Lightning Source LLC
Chambersburg PA
CBHW061251220326
41599CB00028B/5609